P9-CRI-702

Le peintre et son modèle (1920) by André L'Hote demonstrates the relationship between cubism and abstract mathematical ideas of transformations. (CENTRE POMPIDOU, PARIS)

Also by Amir D. Aczel

Chance
Descartes's Secret Notebook
The Mystery of the Aleph
Pendulum
Entanglement
The Riddle of the Compass
God's Equation
Fermat's Last Theorem

The ARTIST and the MATHEMATICIAN

The Story of Nicolas Bourbaki, the Genius Mathematician Who Never Existed

AMIR D. ACZEL

THUNDER'S MOUTH PRESS
NEW YORK

THE ARTIST AND THE MATHEMATICIAN:
The Story of Nicolas Bourbaki, the Genius Mathematician Who Never Existed

Published by
Thunder's Mouth Press
An Imprint of Avalon Publishing Group, Inc.
245 West 17th Street, 11th floor
New York, NY 10011

Georges Braque's *Femme à la guitare* and André L'Hote's *Le peintre et son modèle* appear by arrangement with CNAC/MNAM/Dist. Réunion des Musées Nationaux/Art Resource, NY.

Marcel Duchamp's *Portrait de joueurs d'échecs* appears by arrangement with The Philadelphia Museum of Art/Art Resource, NY.

First printing, June 2006

Library of Congress Cataloging-in-Publication Data is available.

ISBN-10: 1-56025-931-0
ISBN-13: 978-1-56025-931-2

9 8 7 6 5 4 3 2 1

Book design by India Amos, Neuwirth & Associates, Inc.
Printed in the United States of America
Distributed by Publishers Group West

For Miriam, with much love

TABLE OF CONTENTS

ACKNOWLEDGMENTS

I AM HIGHLY INDEBTED to five mathematicians and one artist. They are the mathematicians Pierre Cartier (IHES and ENS, Paris), Barry Mazur (Harvard University), Marina Ville (Jussieu, Paris, and Boston), two mathematicians who insist on anonymity, and the painter Thomas Barron.

I thank Owen Gingerich and the Department of the History of Science at Harvard University for making me a visiting scholar in the department. I am equally indebted to Fred Tauber, director of the Center for Philosophy and History of Science at Boston University, for having me as a research fellow at the center.

I thank my longtime editor and friend, John Oakes, for making this book a reality and for his ever thoughtful editing and advice. Finally, my deepest appreciation to my wife, Debra, without whom this book would never have been written.

The ARTIST and the MATHEMATICIAN

one

THE DISAPPEARANCE

IN AUGUST 1991, Alexandre Grothendieck, widely viewed as the most visionary mathematician of the twentieth century, a man with insight so deep and a mind so penetrating that he has often been compared with Albert Einstein, suddenly burned 25,000 pages of his original mathematical writings. Then, without telling a soul, he left his house and disappeared into the Pyrenees.

Twice during the mid-1990s Grothendieck briefly met with a couple of mathematicians who had discovered his hiding place high in these rugged and heavily wooded mountains separating France from Spain. But soon he severed even these new ties with the outside world and disappeared again into the wilderness. And for ten years now, no one has reported seeing him. His mail keeps piling up uncollected at the mathematics department of the University of Montpellier in southern France, the last academic institution with which he had been associated. The few individuals whom he had once trusted to forward him the select pieces of mail he did want to receive no longer have any way of making contact with him. His children have not heard from him in many years, and two of his relatives who live in southwest France—not far from the Pyrenees—and with whom he had had limited, sporadic contact, have not had a word from him in years. They do not even know whether he is still alive. It seems as if Alexandre Grothendieck has simply vanished off the face of the earth.

During his most active period as a mathematician, from the 1950s to around 1970, a period when he completely reworked important areas of modern mathematics, lectured extensively on his pathbreaking research, organized leading seminars, and interacted with the most important mathematicians from around the world, Alexandre Grothendieck had been closely associated with the work of Nicolas Bourbaki. And some have surmised that Grothendieck's inexplicable disappearance into the Pyrenees was somehow connected with his relationship with Bourbaki.

Nicolas Bourbaki was the greatest mathematician of the twentieth century. Since his appearance on the world stage in the 1930s, and until his declining years as the century drew to a close, Bourbaki has changed the way we think about mathematics and, through it, about the world around us. Nicolas Bourbaki is responsible for the emergence of the "New Math" that swept through American education in the middle of the century as well as the educational systems of other nations; he is credited with the introduction of rigor into mathematics; and he was the originator for the modern concept of a mathematical proof. Furthermore, the many volumes of Bourbaki's published treatise on "the elements of mathematics" form a towering foundation for much of the modern mathematics we do today. It can be said that no working mathematician in the world today is free of the influence of the seminal work of Nicolas Bourbaki.

But what was the nature of the relationship between Bourbaki and Alexandre Grothendieck, and who is Nicolas Bourbaki?

ALEXANDRE GROTHENDIECK WAS born in Berlin on March 28, 1928, to Alexander (Sacha) Shapiro and Johanna (Hanka) Grothendieck. Both parents were ardent and very active anarchists.

Sacha Shapiro was born on October 11, 1889, to a religious Jewish family belonging to the Hassidic movement, in the town of Novozybkov in the region where the three borders of Ukraine, Belarus, and Russia all meet. At times, the town has been Russian, at others, Ukrainian.[1]

Shapiro forsook the Hassidic lifestyle of his parents and was swept with the revolutionary movements of the time. He participated in several uprisings against the Russian tsar, the first one being the abortive 1905 revolution. As a member of an anarchist movement, he took the *nom de guerre* Sacha Piotr. He was first arrested when he was only sixteen years old, when the attempted coup against the tsar failed. Following this arrest, Shapiro was deported to Siberia and spent ten years in jail. Released in 1917, he found himself taking part in the Russian revolution that same year. Again he was arrested, and at some point was condemned to death. He escaped, was recaptured, and ran away again. During his last escape from the Russian police he lost an arm. He then fled from Russia, assuming the false name Alexander Taranoff, and would remain a stateless person for the rest of his life. Soon after his release from prison, before taking flight, he met and married a Jewish woman named Rachel, and with her had a son named Dodek.

But their time together did not last long and, with the aid of woman named Lia, he slipped across the Russian border into Poland. He and Lia then drifted through France, Belgium, and Germany, illegally crossing national borders with unusual agility, and finally arrived in Paris, where they lived together for two years; but Sacha had a string of other relationships with women. Through his anarchist activities he met important figures of world revolutionary movements, including the Americans Alexander Berkman (1870–1936) and Emma Goldman (1869–1940).[2] He moved to Berlin and there, through his involvement with anarchist groups, met Hanka Grothendieck.

Johanna (Hanka) Grothendieck was born in Hamburg on February 23, 1900, to Albert and Anna Grothendieck. Albert had a shoeshine business in town, which at the height of his career he owned; otherwise, he worked there as an employee. Hanka was close to her father, who encouraged her to pursue her gifts of acting and writing.

Soon, the girl rebelled against her parents and her lower-middle-class upbringing and took up with artists and intellectuals. She hoped to become a writer, and early on in her life she worked as an editor and writer for small papers, and also studied to become an actress. She met and worked for German impressionist painters before the urge to travel swept her away from Hamburg and took her to the center of German intellectual life: Berlin.[3] Here she came in contact with artists such as Paul Klee, from whom she bought a small drawing, continued her studies of acting, and became acquainted with members of revolutionary social movements that were active in the German capital at that time.[4]

Taken with social causes, Hanka accepted a position writing for the newspaper *Der Pranger,* which focused its articles on the plight of the disenfranchised. She wrote articles on prostitution and took up the cause of sexually exploited women and girls. Through her involvement with revolutionary causes, she had met a man named Alf Raddatz, and the two were soon married and had a daughter, Maidi. Raddatz could never hold a job, was often away, and did not live with Hanka.

It was during this period of intermittent separation from her husband that she met Sacha Shapiro. He worked as a street photographer, which proved difficult as he had only one arm, but he did it well. The two moved in together. Their son, Alexandre, was born in Berlin on March 28, 1928. Since they were not married, and the mother's married name was Raddatz, Alexandre Grothendieck's name was registered as Alexander Raddatz. Only later, after his mother's divorce from Alf Raddatz, would his name be changed to Grothendieck.[5] Years later, in France, the spelling of his first name would be changed to Alexandre.

For five years, Alexandre Grothendieck lived in Berlin with his parents and his half sister, Maidi. The parents were consumed with their mutual love affair with revolution. Both were angry with authority, which they saw as the source of all evil and misery in the world, and strove to rid society of the ills of government. They truly believed that anarchist movements could bring about redemption from tyranny and freedom for the masses. Poor themselves, they lived on Brunnenstrasse, in a section of the city that was populated by indigents, Jews, and foreigners.[6] Sacha's work as a street photographer brought little income, and Hanka's infrequent jobs writing for the newspaper helped little. Sacha was often away on anarchist missions to France and other countries. Soon, even the little stability the family had enjoyed would be gone.

WITH THE RISE of Nazism in 1933, the situation for the family worsened significantly. Sacha, always on lists of wanted persons and living under a false name and with forged papers, had to escape from Germany. In the middle of the night, without saying good-bye to Alexandre and his sister, Sacha left the family home, crossed the French border, and traveled to Paris. A few months later, when Hanka decided to join him there, Maidi was placed in an institution for handicapped children in Berlin, even though she was not handicapped.

Hanka wanted to place five-year-old Alexandre in a foster home. Some years earlier, she had met an unusual man, a Lutheran pastor named Wilhelm Heydorn, and she knew that he and his wife often took in foster children for a small monthly fee. Hanka wrote a letter to Dagmar Heydorn, Wilhelm's wife. She described her desperate situation, explaining that her "husband," who had only one arm, was a stateless Russian working in Paris as a street photographer and that she was badly needed there to help him earn a living. They were both writing a very important book, she wrote, and this book could not be completed unless she joined him. She said she knew that the Heydorns accepted a small number of foster children, and that she was ready to pay one hundred marks—and added that she could not afford more.

Dagmar's response was encouraging, so in early May 1939, Hanka came to the Heydorn home, bringing the boy the Heydorns would later name "the little Russian." She immediately confessed to Dagmar that she had no money at all and could not pay the hundred marks she had promised, and that there was no great book she and her husband were writing. She merely needed to help her one-armed man earn a living.

Dagmar felt sorry for the boy, who would be facing a very uncertain future if she declined to take him in. She consulted with her husband. Wilhelm Heydorn was a unique individual: he had been a Lutheran priest, an army officer, an elementary school teacher, and a practitioner of alternative medicine. For several years, he had studied to become a doctor, before giving it up for the priesthood. He was an intellectual and the founder of an outlawed political party opposed to the Nazis. Within a few years he would go underground himself, forced to hide from the authorities and live using false papers. Wilhelm agreed immediately with his wife and before they knew it Hanka was gone and little Alexandre was in their care.

For five years young Alexandre was raised by the Heydorns, who had four children of their own and had taken in several foster children whose parents had escaped following the rise of Hitler. From Alexandre's autobiographical notes, we know that the boy was unhappy and deeply missed his parents. He had very little news from his mother during this period, and heard nothing at all from his father. None of his mother's relatives in Hamburg ever came to visit him.

The Heydorn household had strict rules, which was something Alexandre was not used to, having been raised by anarchist parents who did not impose any limitations on his freedom. This made his stay in the foster home even more difficult. Still, he was always appreciative of all that the Heydorns did for him, and always very patient and polite.

Meanwhile, in 1936, Alexandre's parents, ever politically active, went to Spain to take part in the Spanish Civil War. This war was the great hope of all revolutionaries and anarchists living in the continent of Europe. Thousands of anarchists flocked into Spain, coming from as far away as Italy and beyond. They joined the Republican forces fighting the Fascists. But General Franco's fascists were better organized, better trained, and better armed.

The Fascists won the war, and soon the many thousands of foreign soldiers and Republican troops, as well as Spanish anarchists, were crossing the Pyrenees back into France, defeated and demoralized. Sacha and Hanka came back to France disillusioned and depressed. In France, like the thousands of refugees from the Spanish Civil War, the two were viewed as dangerous foreigners and remained under surveillance by the authorities. Sacha resumed his work as a street photographer in Paris, and Hanka found a job teaching young children in the southern city of Nîmes.

In 1939, just before the beginning of the Second World War, the Heydorns came to believe that they no longer could keep a child who looked Jewish, for this was now too dangerous for them. They therefore tried to contact Sacha and Hanka in France. Since their whereabouts were unknown, this was a difficult task; but through the French consulate in Hamburg a message was finally relayed to the parents that they needed to collect their son. Eleven-year-old Alexandre was put alone on a train in Hamburg and arrived in Paris in May 1939, finally to be reunited with his father.[7] Hanka came back from Nîmes and joined the two of them in Paris.

Their time together, however, was to be short. Soon Sacha Shapiro was arrested and sent to the worst French internment camp, Le Vernet, in the Pyrenees. This was a camp in which about 2,000 men were kept in hideous conditions. The inmates were refugees from the Spanish Civil War, revolutionaries, Jews, and other "undesirable foreigners." Life in this camp was extremely difficult, the sanitation appalling, and the inmates constantly hungry. In 1942, the French collaborationist authorities began to deport Jewish inmates to Auschwitz. Sacha was among the first to be deported, and he died in Auschwitz in 1942. It is chilling to read the following matter-of-fact sentence in a report by a French officer to his collaborationist French commanders: "I have the honor to make it known to you that Monsieur Taranoff

. . . was deported on 14 August 1942 in the direction of the concentration camp of Auschwitz."[8]

In 1940, Hanka and her son, Alexandre, were placed in the French internment camp of Rieucros, located in a flat agricultural region in southern France—a place that gets

Grothendieck at the Camp of Rieucros.
(CREDIT: ALEXANDRE GROTHENDIECK)

hot and unpleasant in summer and freezing cold in winter. Conditions in this small camp, however, were better than in other places of confinement in France, and the boy was able to attend school in the nearby town of Mende. Rieucros was a women's camp, and some of the women had children with them. According to Grothendieck's memoirs, he was the oldest of these children, and the only one to study at the lycée in Mende.[9]

Reports by survivors indicate that life at Rieucros was very difficult, and that hunger and deprivations were suffered daily by the poor women who ended up here. At this, or at another camp to which she was later to be sent, Hanka contracted tuberculosis and would eventually die from this disease. But Grothendieck's memoirs do not say much about these hardships. His writings focus on his studies at the local school and the teachers, those he liked and those he cared less about. And he writes about his fascination with words and poetry, and about the magic he found in numbers.[10] Alexandre learned how to rhyme from another boy at the camp, and soon all his sentences rhymed. This was a fun game to play for many hours every day.

Another friend taught him about the existence of negative numbers and games one could play with them. Then he learned how to create and solve crossword puzzles, and this game kept him occupied for days on end in the confinement of the camp. Mostly, he would spend his time alone, and as an adult he would appreciate the gift he received in the camps—the ability to spend time in complete solitude. His lonely hours would teach him how to create thoughts and derive ideas without interacting with a soul.

But clearly life was extremely difficult for the boy and his mother. At the camp were interned "dangerous foreigners":

German Jews, Spanish Anarchists, and Trotskyites.[11] As speakers of German, the Grothendiecks were shunned and harassed by many of the others and certainly by French people in the surrounding villages, since—paradoxically—they were viewed as the enemy rather than as victims of the Germans.

Alexandre grew up in a harsh, confrontational environment in which he was often physically attacked. In order to survive, he developed into a strong fighter and would retain and develop his boxing skills for the rest of his life. His anger at the world he knew escalated to the point that he ran away from the camp with the intention of assassinating Hitler. Fortunately for himself and his mother—since he had no chance of ever completing such a task—he was caught and returned to the camp.

Grothendieck remembers, however, that as the oldest boy in the camp and the only one to attend a high school at a village four or five kilometers away, he had the ability to leave and reenter the camp almost at will.[12] He recalls that he was a good student, but not exceptional. He was already in the habit he would follow throughout his life: of concentrating only on what caught his fancy and completely ignoring the rest. Alexandre did not care what his teachers thought about him. If a topic interested him, he would spend hours on end on it; and if it didn't, he cared about it not at all.

He still remembers, however, the first bad grade he received in mathematics—a field that would become his passion and his career. A teacher had asked him and his schoolmates to prove the "three cases of equality of triangles." Grothendieck's proof was different from the one in the textbook, so the teacher marked him down, even though his proof was "every bit as correct as the one in the book."[13] The teacher, apparently, had such low confidence in his own mathematical abilities that he could not recognize the value of Grothendieck's alternative proof of the theorem. He had to "report to an authority," and that authority was the textbook, Grothendieck lamented.

Vichy France was a difficult place in which to survive for a boy of Jewish origins. And things got worse when the Rieucros camp was closed in early 1942 and its 320 inmates were moved to the concentration camp at Brens, which had stricter rules and even fewer freedoms than Hanka and her son had enjoyed before.

Hanka had heard rumors about a school, supported by a Swiss charity, in the French town of Le Chambon-sur-Lignon. Here Jewish students could study and were hidden from the Nazis. She managed to send Alexandre to this school.[14] Sometime after her separation from her child, Hanka was moved yet again, this time to the camp at Gurs, where she would be kept prisoner until the end of the war.

The following is a short description of Alexandre Grothendieck written by the woman who ran one of the camps at which Alexandre was interned, the camp of La Guespy. It appears to have been written shortly after the end of World War II. M. Steckler in the description below was the camp's surveyor. He used to spend many hours in this camp "ferociously" playing chess with the boy.[15]

> ALEXANDRE GROTHENDIECK
> *Dit Alex le Poète*
> *Allemande, russe?*
> *Mère au camp de Gurs*
> *Enfant très intelligent, toujours plongé dans ses réflexions, ses lectures, écrivant*
> *Très bon joueur d'échecs—parties acharnées avec M. Steckler*

Réclame le silence pour écouter la musique
Sinon enfant tapageur, nerveux, brusque

(Alexandre Grothendieck
called Alex the Poet
German, Russian?
Mother at the camp at Gurs
Very intelligent child, always plunged deep in his thoughts, his reading, a writer
Very good chess player—chess matches set against M. Steckler
Demands silence for listening to music
Otherwise, a loud, brusque, nervous child.)

The mountain resort town of Le Chambon-sur-Lignon, in a wooded area at an elevation of 3,000 feet in central France south of Saint Etienne, had been transformed into a stronghold of the French resistance and became a haven for the few Jews and other persecuted people who could find their way here. Virtually the entire population of this town actively hid Jews from their Nazi persecutors. The force behind this amazing countercurrent in the generally anti-Semitic and antiforeign atmosphere of wartime France was one man: the Protestant pastor André Trocmé (1901–1971).

Trocmé was a Huguenot, and was born in northern France near the Belgian border. He studied for a time in the United States, and in New York met Magda Grilli, of Italian and Russian origins, who would become his wife. When he was sent to be the parish priest of Le Chambon, he began to preach tolerance, and his teachings were well received by the local Protestant population, which had been attuned to rebellion against authority ever since the revocation of the Edict of

Nantes in 1684.[16] These people had a history of siding with the persecuted, and saving those the authorities were hunting down for deportation and execution. The people of Le Chambon-sur-Lignon literally risked their own lives daily in order to save the Jews living and hiding among them.

But life continued to be dangerous, and there were frequent roundups of Jews by the Nazis. Grothendieck later described how he often had to escape into the woods every time the Nazis were approaching, and hide out for several days at a time, with little or no food or water.[17]

Alexandre attended the local school, the Collège Cévenol, from which he eventually earned his baccalauréat, which would entitle him to enroll at a French university. His education in Le Chambon, and earlier at Mende, was spotty at best and lacked both continuity and depth. But the boy had a strong ambition, and he had a special affinity for mathematics. In addition to the fact that the teaching at these small schools in wartime France was not good, the textbooks were inadequate.

The young student found the problems in the mathematics textbooks so repetitive and trite that he stopped using them. What bothered him the most, however, was the fact that the problems in the books appeared as if out of thin air—with no reason or motivation behind them. He felt that these problems did not illuminate the material but rather were arbitrary and senseless. He therefore made up his own problems, and then spent many hours solving a problem that interested him, ignoring everything else.

What most concerned Grothendieck was the fact that none of his mathematics texts at the Collège Cévenol gave a good definition of length, area, and volume. Thus, at a high school in a village near a concentration camp, a young boy was concerned with mathematical problems that were far above the level and the place and that were, indeed, not "pulled out of thin air" as the problems in the textbooks seemed, but had an important grounding in the real world. The boy Alexandre Grothendieck was interested in the *theory of measure*, even though he could not have known it by that name at this time.

Grothendieck wanted to be able to find the length of a curve, the area of a triangle with given sides, and the volume of a regular solid with a given edge.[18] The problems of measure theory would continue to occupy his mind after the war, when he was a student at a university. He would re-derive on his own the theory of measure, which, unbeknownst to him, had been developed a few decades earlier.

The Pyrenees about twenty kilometers south of Grothendieck's hideout. (AMIR D. ACZEL)

WHILE ALEXANDRE AND his mother were doing everything they could in order to survive in the camps of southern France, a number of French mathematicians were also going through the upheavals of the war, in other parts of the country. The war had completely disrupted the state of French mathematics, just when it was beginning to make progress—as mathematicians started to reorganize following the disastrous outcome of the First World War, which had decimated half the graduating classes of French universities of the years 1910 to 1916.[19]

Some of these mathematicians, who had begun to rebuild French mathematics just a few short years before the outbreak of the Second World War, were now refugees themselves. As professors they had been fired from their positions, or for other reasons found themselves without work and with the need to hide from the authorities. French mathematics revived for a short time during the period between the two wars, and then was fatally disrupted by the onset of the Second World War. French mathematicians were now struggling to survive as the world entered its most insane period in history.

two

AN ARREST IN FINLAND

ON THE EVE of the outbreak of World War II in the summer of 1939, as fear and chaos were sweeping Europe, French mathematician André Weil was spending a peaceful and serene time with his wife, Eveline, in the bucolic setting of a villa on the island of Lökö on the Gulf of Finland. The Weils were the guests of Lars Ahlfors, the renowned Finnish mathematician, and his wife, Erna. Years later, Weil described their stay in his memoirs as follows:[1]

> Our visit with the Ahlfors was a time of unadulterated serenity. . . . Our small island was easily explored: besides our villa, there was only a small farm with four or five cows, some distance away. It was the season of white nights, close to midsummer night's eve. The air was invariably pure and clear, transparent beyond words.

Their days were spent walking on the island, going on long outings by boat to explore nearby islands, and gathering at ten or eleven in the evening in the dining room of the villa for tea and sandwiches by the light of the setting sun. As Weil described it, "We felt we were somewhere outside of time."[2]

After a few weeks with their hosts, André and Eveline Weil traveled farther in Finland to stay at a small hotel on the shore of Lake Salla, close to the Russian border.

> Our days were calm and uneventful. We spent many hours on the lake in the hotel skiff, or sitting at the water's edge. I had with me my faithful typewriter, and there I typed the outline for a report. . . .[3]

Weil also dictated long passages to his wife. All that time, they were taking pauses, looking at the beautiful surroundings, the hills near the Russian border. Unbeknownst to them, some people at the hotel became suspicious of a couple sitting outside, surveying these border surroundings and typing incessantly. They came to the conclusion that the two were Russian spies and reported their suspicions to the Finnish police, which opened a file on André Weil.

The pair traveled farther north and reached the Arctic Ocean, where André even tried to swim for a few seconds in the icy water. Then they continued along the Finnish border with Russia, and came to Helsinki, arriving there just before the Second World War broke out. But Eveline had to return to France. She had a five-year-old son from a previous marriage whom she had left with her mother and she needed to get back to take care of him. André knew, however, that if he returned to France he would immediately be arrested. He had to stay away, and he had brought with him a large amount of cash, in U.S. dollars, to support this stay.

They were sitting in an outdoor café on the Esplanade when they heard of the declaration of war, and they felt as if they had just lost a beloved friend.[4] Eveline decided to stay a bit longer with her husband, whom she felt she might never see again. Ahlfors helped them find a small room with the use of a kitchen in a house not far from his own house in the city. And as the hostilities of war raged south of them on the main part of the continent of Europe, the pair took walks in the parks along the Baltic Sea, feeding the tame squirrels that came and sat in their hands.

But by October 20, 1939, Eveline could wait no longer. Communications with France were in jeopardy of being discontinued at any moment, and she left by train to Sweden, going from there through Denmark and Holland back to France, where she was reunited with her son and her mother in a small town in the countryside.

André Weil was astute enough to understand that he was not safe in Finland. He knew that the Russians were planning to attack, and his intuition told him that Sweden was probably his safest bet. But because the war had begun, a visa would have been needed for travel to Sweden. Obtaining a visa would have required the cooperation of the French authorities—something Weil could not hope for in his case. So he stayed on in Helsinki, in his little room, and continued his work.

On November 30, the Russians began their bombardment of Helsinki. Weil saw all his neighbors rush out and leave for the countryside. He followed them. Around noon, the air raid alert ended, and people began to return home. He walked toward a nearby square, to see what was happening. But his foreign clothing and his myopic squint made him stand out in the crowd. A policeman approached him, and proceeded to arrest him.

Weil was taken to the local police station. When the Finnish police headquarters was contacted about the arrest, it was discovered that there was already a file on André Weil for suspicion of spying. He was immediately transferred to the central police station in Helsinki, where he was interrogated for several hours. Then the police conducted a thorough search of his room, where they found what they believed was condemning evidence:[5]

> The manuscripts they found appeared suspicious. . . . There was also a letter in Russian, from Pontrjagin, I believe, in response to a letter I had written at the beginning of the summer regarding a possible visit to Leningrad; and a packet of calling cards belonging to Nicolas Bourbaki, member of the Royal Academy of Poldevia, and even some copies of his daughter Betti Bourbaki's wedding invitation.

There followed a "rather calm, if lengthy, interrogation at the police station." The interrogation was conducted in German, since the police officer in charge knew that language and so did Weil. Weil was aware that the policeman was trying hard to catch him contradicting himself in his statements. At some point the policeman exclaimed: *"Sie haben gelügt!"* ("You lied!"—but with a grammatical error). Weil's immediate answer was "*Nicht* gelügt, *man sagt* gelogen" ("it's not *gelügt,* one says *gelogen"*). The policeman did not seem offended by this response, and stood corrected.

Unbeknownst to Weil, the Finnish authorities had concluded definitively that he was a Russian spy, and they condemned him to death. On December 3 or 4, a Finnish official named Rolf Nevanlinna was present at a state dinner also attended by the chief of police. When coffee was served, the latter approached Nevanlinna and said, "Tomorrow we are executing a spy who claims to know you. Ordinarily I wouldn't have troubled you with such trivia, but since we are both here anyway, I am glad to have the opportunity to consult you." Nevanlinna asked him, "What is his name?" The chief of police answered, "André Weil."[6]

ANDRÉ WEIL WAS born in Paris in 1906. His paternal grandfather was Abraham Weill, an important member of the Jewish community in Strasbourg. As such, he was often called upon to settle disputes between members of the community. When his wife died, he married her sister, as dictated by Jewish tradition. With her, he had two sons, André's father, Bernard Weil, and his uncle, Oscar. At some point in his life, Bernard shortened his family name, losing the second *l*.

By law, the people of Alsace could choose either German or French citizenship.Weil's parents and their respective families both chose France as their homeland and moved to Paris. Bernard Weil studied medicine, and in Paris he became a well-respected general practitioner known for the reliability of his diagnoses and the excellent treatment he provided his patients.[7]

André Weil's mother, Selma Reinherz, was born in 1879 in the Russian city of Rostov-on-the-Don to Austrian Jews who had immigrated there. Her father was a prosperous grain merchant. Three years after she was born, following the pogroms of 1882, the family moved to Paris.

Selma and Bernard were married in 1905. Theirs was an arranged marriage, but in time they grew to love each other. Her family's wealth helped Bernard set up his clinic in Paris and enabled him to embark on a very successful medical career. They had two children, André and Simone (who would become a famous philosopher). The family lived in an apartment on Boulevard Saint Michel in the Latin Quarter in Paris, by the Luxembourg Gardens. Every day, Selma would take Simone and André for a stroll in these beautiful gardens and the children would run and play and enjoy the best of a Parisian child's life. André Weil was growing up in a world of privilege and his enviable circumstances stand in sharp contrast with those of the formative years in the life of Alexandre Grothendieck. Both men would reach great heights in mathematics, but Grothendieck's contributions to this field would ultimately be valued as greater than those of Weil. Yet, while Weil would be spared no advantage in his education and exposure to great ideas, Grothendieck would enjoy nothing of the sort and would have to develop his immensely important ideas in an intellectual vacuum while barely surviving physically.

Early on in his school days, André Weil was recognized for his amazing mathematical abilities. He learned to read between the ages of four and five, and soon he began to read books that were far above his level in school.[8] He also started to read books on mathematics, which he enjoyed very much.

When the First World War erupted in 1914, Dr. Bernard Weil was called to serve his country in the medical corps and was assigned to various places. After a few months he became ill himself and in 1915 was sent to recuperate in a hospital in Menton, in the south of France on the Mediterranean coast. The family followed him from Paris and stayed for some time in this region. It was during this period, when he was nine years old, that André Weil became aware of his great mathematical abilities. His parents bought him a subscription to a mathematical journal for teachers, the *Journal de Mathématiques Elémentaires*. In this journal were published problems to be solved by the teachers, and those who solved them got to have their names in print. After he tried his hand at these problems, André Weil's name appeared on the pages of almost every issue of the journal. The boy also developed a taste for grammar and literature, and especially enjoyed a French grammar class in which the teacher created an algebraic method of notation. This early experience with mathematical concepts applied to language inspired him and would exert an impact on his development of abstract mathematical concepts in his future mathematical career.

In 1917, after he attended various schools at the locations to which his father had been assigned—Paris, Chartres, and Laval—André Weil taught himself enough Greek and mathematics to gain acceptance to the classical section of a higher grade at school. He read the *Iliad*, and then devoured a book on Indo-European linguistics, which whetted his appetite to study Sanskrit. He was well on his way to becoming a Renaissance man.

But Weil's great breakthrough in understanding and loving mathematics came a year later, after the family had moved back to Paris and he was accepted at the prestigious Lycée Saint-Louis, one of the best high schools in France. Here, his mathematics teacher was a man named M. Collin, a teacher with unusual abilities and teaching habits. Once, M. Collin had been assigned to a very difficult class, a class known to be comprised of troublemakers, at the preparatory school in Saint-Cyr. One day, he found out that the students in this class had planned to put him to the test. He entered the classroom, sat down, and spent the entire period staring at his students. He never had trouble with these students afterward.

In mathematics, M. Collin taught the important concept of rigor. André Weil wrote in his memoirs that Collin "showed me once and for all that mathematics operated by means of rigorously defined concepts." Once the definition of a function was given, for example, the teacher did not tolerate anyone using the word "function" for anything that did not correspond precisely to the given definition.[9] This was a very important lesson, which started Weil in a direction in mathematics that would exert a strong impact on the development of the discipline throughout the entire century and up to the present time. Mathematics had by that time often been pursued in imprecise, even vague, ways. Newton, for example, had defined the derivative in calculus using the vague notion of a "fluxion," which had not been clearly and rigorously defined. And the same was true for later mathematicians. Weil would grow up to become one of the greatest proponents of clarity, precision, and rigor in mathematics.

Weil progressed quickly and soon took his baccalaureat examinations and enrolled in the required preparatory classes for one of France's "Great Schools," the École Normale Supérieure (ENS). There, he was introduced to one of the greatest French mathematicians of the time, Jacques Hadamard. Later, Hadamard would become Weil's doctoral dissertation advisor.

Even though he was getting on in years, Hadamard retained an amazing clarity of mind and an incredible sharpness in his mathematical work. He was also unusually modest and kind. He welcomed Weil into his fold. Through Hadamard, the young Weil came in contact with virtually all of France's great mathematicians, including Henri Lebesgue and Elie Cartan. The student benefited from taking courses from these renowned professors, and from attending many lectures and symposia.

In 1921, Einstein's theory of relativity was all the rage in Europe. This was three years after Arthur Eddington provided definitive physical proof for the general theory of relativity by observing the bending of starlight around the sun during an eclipse expedition to the island of Principe in the Atlantic Ocean in 1918—exactly as predicted by Einstein's theory three years earlier.[10] Fifteen-year-old André Weil read Eddington's book about relativity, understood both the physics and the complicated mathematics Einstein had used, and explained it to his parents during a vacation in the Black Forest.

The following year, Albert Einstein was to give a talk about relativity at the Collège de France. The most important French mathematicians and scientists, as well as members of French high society, were going to attend, and admittance to this monumental event was by invitation only. Jacques Hadamard arranged for the young Weil to be invited to this talk.

The precocious young student then turned his attention to the study of Sanskrit. He had read Indian poetry in translation and was so taken with it that he decided he had to read it in the original. Within a short time, Weil became proficient enough in Sanskrit to be able to read the *Bhagavad Gita*. Throughout his life he would often quote from the *Gita* and derive from it both artistic pleasure and spiritual guidance in life and its travails. At the same time, he also became enamored with ancient Greek and Latin books. The boy would save his money to buy books printed in the 1500s. His favorite was a 1560 edition of the *Iliad* in Greek and Latin, which would become his inseparable companion for many years. He also bought Aldus's *Plato* printed in 1513. Weil found some of these old books in the boxes of booksellers along the banks of the Seine, as one can still do nowadays—although, unlike today, he was able to purchase them at reasonable prices.

German mathematics had survived World War I, and Germany had thus retained its hegemony over mathematics, which it had acquired sixty years earlier.[11] Mathematics was well-organized in Germany, with many fine mathematicians at the universities of Berlin, Göttingen, and elsewhere. The Germans were leaders in analysis, algebra, and other areas of modern mathematics. They had a tradition of science and technology, and within this tradition, mathematics was very important, a subject to be taken seriously, to be studied and developed. German mathematicians were the unchallenged rulers of the discipline worldwide.

André Weil was proficient in German because of his family roots in Alsace and his grandparents' Austrian and German origins. Both his parents spoke German and actually used it as a secret language when they did not want André and Simone to understand what they were saying. This was strong enough incentive for both children to learn this language well. André became interested in German mathematics. He read German papers in the original and studied the advances the Germans were making in this field. As an adult, he would be influential in France's reassuming a leadership role in mathematics vis-à-vis Germany.

As a student at the École Normale in Paris, Weil often checked out German books on mathematics from the library and learned from them about the methods and ideas that the German mathematicians had been using. This was an excellent source of knowledge not used by many of his peers. One of the German mathematicians whose work especially interested Weil was Bernhard Riemann, a man who in the nineteenth century rewrote much of geometry and topology and made important contributions to other areas such as number theory. Equally important, Weil attended Jacques

Hadamard's seminar at the Collège de France, learning much high level mathematics—material that was well beyond the usual courses at the École. "The *bibli* (library) and Hadamard's seminar, that year and the following ones, are what made a mathematician out of me," Weil wrote in his memoirs.[12]

Riemann was one of the mathematicians Weil considered a "great mind," and, through reading him and the Greek poets, Weil came to the conclusion that "what really counts in the history of humanity are the truly great minds, and that the only way to get to know these minds is through direct contact with their works."[13] According to Weil, his sister, the philosopher Simone Weil, also came to a similar conclusion, perhaps through his influence.

Weil was also impressed by a phrase he read in the work of the great French mathematician of the late nineteenth and early twentieth centuries, Henri Poincare, whom Weil quotes as saying, "The value of civilizations lies only in their sciences and arts." With such ideas in his mind, Weil wrote, he had no choice but to "dive headlong into the works of the great mathematicians of the past."[14]

Weil studied the works of Riemann, and the writings of the French mathematician Camille Jordan helped him understand Riemann's concise writings. He also read Felix Klein's mimeographed lecture notes, which further helped him flesh out the ideas of Riemann. For a beginning university student, Weil was moving very deep and very fast into complex contemporary mathematics.

However, living in such a specialized intellectual environment, he did not want to leave behind his interests in Sanskrit, Greek and Roman poets, and music. With the aid of the English translation of the *Sacred Books of the East*, Weil sat in an overgrown country house his parents had built in Chevreuse and read the *Bhagavad Gita* from cover to cover. "The beauty of the poem affected me instantly," he would later write, "I felt I found in it the only form of religious thought that could satisfy my mind."[15] Judaism never meant much to Weil (or to his sister, who converted to Catholicism). He claimed not to have known that he was Jewish until the age of ten or twelve, and, once he found out about it, not to attach to this fact any great importance.[16]

After his graduation in 1925, Weil began to travel. He had always been fascinated by Italian art and he longed to see the great masterpieces, so he set out for Rome and spent some time there as well as in other Italian places: Naples, Ravello, and Sicily. In Italy, he met some of the best mathematicians that country could boast, including Volterra, who hosted him at his house for some time. He also met foreign mathematicians who were visiting Rome: Mandelbrojt and Zariski.

Then Weil applied for and obtained a Rockefeller Foundation grant allowing him to visit Germany. Finally, he had the opportunity—the rare privilege, considering how young and unknown he still was—to meet some of the best German mathematicians of the day. He spent time in Göttingen with the mathematician Courant, who would later immigrate to the United States. And in Berlin, he met the renowned master of mathematical foundations, L. E. J. Brouwer. Then, continuing his whirlwind tour of Europe, he went up to Stockholm, where he met Mittag-Lefler, the foremost Swedish mathematician of all time and the founder of an important mathematical journal, for whom Weil had agreed to do some mathematical work. He stayed at Mitag-Lefler's own villa, admiring his books and works, and hearing from him many stories about his life in mathematics.

Mitag-Lefler would begin every day by speaking French

with Weil, and when he got tired he would switch to German, then when he got even more tired, he would switch to Swedish. Finally, he would remember: "But I forgot you don't know Swedish. We will continue our work another time." After two weeks, according to Weil, he understood enough Swedish to follow a conversation of this kind.[17]

Returning from Stockholm to Göttingen by way of Copenhagen, he took a commercial flight from Copenhagen to Lübeck, and from there he returned to Göttingen through Hamburg. Weil notes in his memoirs that this form of travel was still very uncommon. The young man liked it very much and decided to combine such travel with doing mathematics. He thus became the forerunner of modern mathematical and scientific researchers, who conduct much of their work nowadays through frequent travel to professional meetings worldwide. At this time, Weil also resolved to become a "universal mathematician" in the likeness of his mentor, Hadamard—someone who knows "more than non-specialists and less than specialists about every mathematical topic."[18]

Back in Göttingen, André Weil was taken with the workings of the tightly knit mathematical group headed by Max Dehn. Here, he realized that mathematics can be pursued most effectively—and with much enjoyment—when done together in a well-organized group of highly motivated and gifted people. This idea would stay with him and guide him in his career.

Weil now turned his mathematical attention to Diophantine equations—a topic that had always interested him—and conducted work in this field, including an analysis of Mordell's conjecture, an immensely powerful conjectural statement linking two disparate areas of mathematics: topology and the theory of numbers. Weil would continue the work begun on this conjecture upon his return to Paris. In Paris, topics in this general area within number theory would form the basis of Weil's doctoral dissertation, written under the guidance of Hadamard.

While in Germany, Weil became interested in modern art. He visited a number of exhibitions in his travels, including one of the works of Nolde, which he saw in Hamburg, and was influenced by the new ideas of painters such as Picasso on form and shape and structure. Once, in Zurich, Weil visited an exhibition of Picasso. "Rightly or wrongly," he wrote, "it seemed to me that his art juxtaposed the profoundly serious with the prankish—a mixture that was not without charm for the *Normalien* [a person from the École Normale Supérieure] I still had in me. We mathematicians were only vaguely aware of what was called 'the crisis.'"[19]

Modern art broke form and deliberately destroyed all norms that had existed before it, in a way that was not dissimilar to the way Einstein, in constructing his theories of relativity, had destroyed the supremacy of classical physics. These ideas would affect Weil's intellectual development—and his approach to mathematics. Mathematicians were already breaking away from past strictures in their field to forge ahead into new territory.

While in Germany, Weil had read the manuscript of a mathematical paper in German written by Dehn. At the end of the paper, which was to be published in the Proceedings of the Berlin Academy, Weil found the following sentence (here translated into English), which was to be deleted from the published version of the paper: "It is a bourgeois, who still does algebra! Long live the unrestricted individuality of transcendental numbers!"[20] Dehn's dictum represented ideas that were already in the air, so to speak, in the first few decades of the twentieth century—ideas that reflected what was happening within the general culture: that the past should be left behind and that new ideas should be pursued vigorously. This trend was already apparent in art, literature, architecture, and, of course, physics. The time had come, these visionary mathematicians felt, to bring about similar changes in mathematics.

While still in Germany, Weil made his first important mathematical discovery, involving Mordell's calculations on elliptic curves. Mordell's conjecture linked number theory with a particular topological invariant: the genus of a surface. The genus of a surface is the number of holes in it. Thus a doughnut has genus 1, a coffee cup with two handles has genus two, and a sphere has genus zero. Mordell's conjecture was a statement about the genus of a particular surface derived from various equations studied by number theorists.

Weil was able to extend the calculations of Mordell to curves of genus greater than 1. This was a step in the direction of proving this immensely difficult conjecture. Weil took his new results to famous algebraist Emmy Noether and her group in Göttingen, trying to interest them in his work. But he noted with much regret that their attention lay elsewhere: on non-commutative algebras.[21]

At that time, Weil's results were preliminary, and it would take him more than a year to complete them, work that would constitute his doctoral dissertation. When he came to present it to Hadamard, the latter suggested that he try to extend the results to the point of proving Mordell's actual conjecture. But Weil understood that this conjecture was too complicated to prove at that time. In fact, it would take over half a century before this would be accomplished by the German mathematician Gerd Faltings. Weil's results were written up, and he was conferred his doctoral degree. Then he was ready for more travel, this time to the land he had been dreaming of since he first started to read the *Bhagavad Gita*—India.

But before this would be possible, Weil had to fulfill his obligation to do his military service. While most graduates of the École Normale in scientific areas served as officers in the artillery after a short training program, Weil had not had this required training while he was a student. Special arrangements were made, therefore, for him to serve as an officer in the infantry. He served what he called a "peaceful year" in the infantry, during which he slept most nights at home in Paris. He managed to shorten this service by asking for free time to correct the page proofs of his thesis, and otherwise inventing excuses—academic or other—to be away from his unit. Military life did not agree with the young man and he was eager to put this year behind him and to move on.

In 1929, an opportunity materialized for Weil to go to India. He wanted to go there so badly that he even agreed to teach French civilization rather than mathematics when it seemed that a math position would not be available. But at the last moment, a mathematics position did materialize for him at the Aligarh Muslim University. Weil then booked passage on an Italian ship of the Lloyd Triestino Line, and enjoyed his voyage and his visits to several cities in India before beginning his work. At age 23, he became a department chair at this Indian university—something that could hardly be contemplated today. This appointment demonstrates the length to which Indian universities were willing to go during the early part of the twentieth century in order to modernize and westernize their teaching and research.

Thus André Weil, with a doctorate barely in hand, found himself responsible for completely revamping the mathematics teaching at the Aligarh Muslim University, a responsibility that gave him the power to recommend firing and hiring of faculty members.

The university was organized in the British model of higher education, in which the department had one professor, in this case André Weil; it also had a reader, and two lecturers. Weil eventually fired the two lecturers, and kept the reader—even though he felt that none of the three had any great skill or qualifications ("At the age of twenty-three, therefore, I held the fate of these three pathetic characters in my hands")[22]

For one of the newly vacated reader's positions, Weil hired a man named Vijayaraghavan, who had been a student of famous mathematician Hardy in England but had not earned a degree. Having met Vijayaraghavan, Weil regretted not having fired the reader as well, since that would have allowed him to hire the gifted Vijayaraghavan at a higher-level position. Other than hirings and firings, André Weil enjoyed India, traveling widely, from Madras in the south to Kashmir in the north, rereading the *Gita* all along the way. But two years after Weil arrived in India, Vijayaraghavan resigned his position and quickly moved to another school.

When Weil caught up with him, Vijayaraghavan explained that the administration of the Aligarh University had told him that he would be promoted to Weil's position—after they fired the latter. Vijayaraghavan did not want to play any part in this drama, even indirectly, and therefore resigned. While Weil was writing his own letter of resignation, having heard this story and understanding that he would soon be fired, he was served with the document ending his tenure at the university. He then returned home to France.

While still in India, in 1931, Weil decided to spend his summer vacation from the Aligarh Muslim University back home in France, as well as to pay visits to Göttingen, Leipzig, and Berlin. While Weil was in Paris, Bruno Walter was directing the complete series of Mozart's great operas. Places at the theater to see these performances were at a premium, and there was no way for the young man to obtain a ticket. But he would not give up, as he wanted very badly to hear *The Magic Flute*. He put on the Nehru tunic he had bought in India and passed himself off as an Indian dignitary who had traveled to Paris for the express purpose of seeing this opera. "I made such an impression," he wrote, "that I obtained satisfaction."[23]

It was time to get a "permanent" teaching position—not that Weil cared about one. He was much more interested in travel, and his bedside companion was the master schedule of all international trains in Europe. But an acquaintance Weil met on the street in Paris told him that he should start to think seriously about a position—for the pension it would afford for his waning years. Somewhat reluctantly, Weil took a position at the University of Marseille. He was so uninterested in it, however, that when the dean of the school asked him to come down to Marseille before his appointment day, he refused and said he would come right on the day he was to start his teaching.

Shortly afterward, at the start of the academic year in November 1933, Weil joined the faculty of the University of Strasbourg. (His friend, Henri Cartan, was there with him.) Weil would stay on the faculty until 1939, but trouble started for him because he had been an infantry officer, and the French nation expected him to serve as a reserve officer as the Second World War was about to begin. But Weil had other ideas and he fled to Finland.

WHEN ROLF NEVANLINNA was told by the chief of the Finnish police that the spy to be executed was André Weil, he told the chief, "Yes, I do know him. Is it really necessary to execute him?" The head of police replied, "Well, what do you want us to do with him?" Nervanlinna responded, "Couldn't you just escort him to the border and deport him?" To this the chief replied, "Well, there's an idea. I hadn't thought of it."[24] So it was that Weil was placed in a locked compartment in a train heading north with three other inmates. They arrived at the Gulf of Bothnia near the Arctic Circle, and Weil was made to walk across a bridge to the other side, where Swedish border guards promptly arrested him. After several weeks of being dragged from one location to another, he was transported by ship via Britain to France, where French military police took custody of him.

On Friday, May 3, 1940, André Weil, a lieutenant in the French Army, faced prosecution for desertion in a legal process conducted inside the French military prison in Rouen, Normandy. After hearing the prosecution's case against him, the defendant answered, "If I had to don a uniform, I am ready to do whatever the army asks me to do." The judge sentenced him to serve five years in the Rouen prison and to be demoted to private.[25]

The day after his conviction, Weil's sentence was suspended in a deal worked out to allow him instead to volunteer to serve the army in a combat unit, and he was summarily transferred to the front. A few days later, he was sent to his new posting with an infantry unit in Cherbourg. A month later, German troops stormed Rouen and all the prisoners that remained in this military detention facility were shot by their guards as the latter fled. Even at that time, however, French authorities were still investigating the case of André Weil and the suspicious documents the Finnish police had discovered in his apartment.

three

ESCAPE

AS WEIL WAS traveling to his new posting in Cherbourg, now as a private in the French army, another French mathematician was fleeing Grenoble for the tiny village of Saint-Pierre-de-Paladru.

Laurent Schwartz was born in Paris on March 5, 1915. His father, Anselme Schwartz, was born in a small town in Alsace in 1872, soon after the region was annexed by Germany. As Anselme was a patriotic teenager, he immigrated back to France at fourteen. He eventually studied medicine and in 1907 became the first Jewish surgeon employed by a Paris hospital. Anselme married a cousin, Claire Debré, who came from a prominent French Jewish family, among whose members could be counted the chief rabbi of Neuilly, a president of the national academy of medicine, and a number of Gaulist politicians.

Laurent had two brothers, and his great uncle was Jacques Hadamard, one of the most famous French mathematicians of the late nineteenth and early twentieth centuries, and André Weil's mentor. After graduation from high school, Laurent had to choose between classics and mathematics (he was very good in both subjects) and he chose mathematics.

After the grueling preparatory courses, Laurent Schwartz was admitted to the École Normale Supérieure in Paris. At these preparatory courses, called *hypotaupe* and *taupe*—French for "submole" and "mole"—Laurent met a pretty young woman, Marie-Hélène Lévy, who was also working hard to get into the mathematics program at the ENS. She was the daughter of the preeminent French mathematician, Paul Lévy, who pioneered the modern theory of probability. Lévy became Schwartz's academic advisor at the ENS. But soon Lévy also became his father-in-law when Laurent and Marie-Hélène married.[1]

After graduating in 1937, Schwartz enlisted in the military, in order to put behind him this national obligation as quickly as possible so he could begin his life of teaching and research in mathematics. During this time, Schwartz also became politicized: he was concerned with the lives of colonized peoples in Africa and elsewhere. He became an anti-Stalinist, and espoused the thinking of European leftist groups that were fashionable during this period. After his discharge from the army, Laurent returned to Paris and continued to work on mathematics. But as war broke out in 1939 it became difficult for the Schwartz family to continue living in France—and especially so after the occupation. Surprisingly, the fact that Laurent was Jewish escaped the authorities, and he continued to be paid his national research salary. In 1942, this funding source dried up for him, but he continued to receive a small stipend through the Michelin foundation.

On a visit to Toulouse, Laurent and his wife met Henri Cartan and Jean Delsarte, who told them that several French mathematicians who were also caught up in the upheaval of war found positions at the University of Clermont-Ferrand, in the central-southern section of the country. At Clermont-Ferrand, where the Schwartzes subsequently found employment, they met other important French mathematicians, including Jean Dieudonné, Charles Ehresmann, and Szolem Mandelbrojt. All of them were associated with Nicolas Bourbaki, and through them the Schwartzes became acquainted with Bourbaki. It was during this time that Schwartz finished his doctoral dissertation on the approximation of real continuous functions by sums of exponentials.

But then in 1943, things became extremely problematic for Laurent Schwartz and his family. The Nazis invaded southern France and the Schwartzes had to procure false identities to avoid being deported to the death camps. Their child was born during this difficult period and had to be hidden in the hospital when the Nazis came to look for the Jewish baby, so the young family had to flee as quickly as possible.

But while Paul Lévy and Marie-Hélène were safely out of the grip of the Nazis, Schwartz had been separated from them and remained trapped in Grenoble, where he had by then taken the position of professor. He had to make a quick decision. Schwartz moved secretly to Saint-Pierre-de-Paladru, where his wife later joined him.

But Laurent needed a source of income, and found that there was need for mathematics tutors in schools in the area. He traveled from village to village several times a week teaching mathematics. As the Germans advanced, however, his travel was becoming more and more dangerous. Schwartz had false papers that gave him a non-Jewish identity, with a false name he had chosen. By signing his name—Schwartz—and tracing over the letters, the *S* remained intact, the *c* was now an *e,* the *h* and the *w,* when traced over, became an *l* followed by an *i* and an *m,* the *a* remained the same, as did the *r* and the *t,* and the final *z* became an *i* and an *n.* At least that was how his handwriting looked to him. The final outcome spelled Selimartin. So this became Schwartz's new name, and he was able to obtain false papers bearing this new name that did not sound Jewish.[2]

But Schwartz knew that the Nazis were making men they caught in their roundups undress, and were deporting to the camps everyone found to have been circumcised. He knew he had to be extremely careful not to be caught in such a roundup.

One night, Schwartz missed the last train home and had to spend the night in the village at which he taught that day. There were two hotels in the village. One was comfortable, and heated. The other had bare furniture and no heating. It was in the dead of winter, and this night was especially cold. But his instinct told him to check into the bad hotel. Schwartz spent the night awake in his uncomfortable bed, covering himself with the blanket, his coat, and other clothing, but still he was too cold to sleep. He spent the night reading a mathematical manuscript. Around midnight, the Nazis entered the other hotel, where they found two Jews and sent them to their eventual deaths. The soldiers never came to Schwartz's hotel and, in the morning, he returned to his home.[3] The manuscript Schwartz spent that night reading was authored by someone he knew: Nicolas Bourbaki.

BACK IN PARIS, two mathematicians, the father and son Élie Cartan and Henri Cartan, were spending the war years in relative safety. They were not Jews, nor members of any other persecuted group. The father, who was a prominent French scholar, had traveled to Rouen to help André Weil in his trial. But he was never even asked to sit down during the process, and his words in defense of his son's good friend were all but ignored. Soon after he returned home, the two men received a letter from Weil, and were happy to learn that his sentence had been commuted. Attached to the letter was a mathematical paper. Both father and son knew its author well.

Henri Cartan was born in Nancy, France, on July 8, 1904. His father, Élie, was a professor of mathematics at the University of Nancy. In 1909, Élie Cartan was appointed to a professorship at the Sorbonne, and the family moved to Paris. The younger Cartan attended schools in Paris and in Versailles, and in 1923 he was admitted to the École Normale Supérieure in Paris. Three years later, upon graduating from the École, Henri Cartan won a scholarship to allow him to pursue work on his doctoral dissertation. When he completed this work and received his doctorate in mathematics in 1928, he obtained a teaching position at a high school in Caen in Normandy.

After a year he moved to the University of Strasbourg, where he obtained an entry-level faculty position in mathematics. He then taught at Lille for two years but returned to his teaching position at the University of Strasbourg and remained there until the war broke out. In September 1939, the entire University of Strasbourg moved to Clermot-Ferrand in central France to try to escape the dangers of war. It was here that Cartan would again meet André Weil as well as Laurent Schwartz and other French mathematicians trying to stay out of harm's way.

CLAUDE CHEVALLEY WAS born on February 11, 1909, in Johannesburg, South Africa, where his father, a French diplomat, was the Consul General of France. Back in France, the boy grew up and attended school at Chançay, where his parents had bought a sprawling property in 1910. Claude attended high school in Paris, and at age seventeen was admitted to the École Normale Supérieure, where he met and befriended André Weil, who had just returned to Paris from his travels in Italy and Germany. The two shared a passion for number theory, especially algebraic numbers, which was not an important topic in France at that time, but one that the two friends had hoped to bring into prominence.

Chevalley, like Weil, traveled to and stayed in Germany and was influenced by the achievements of German mathematicians—especially Emil Artin in Hamburg, under whose tutelage he wrote his dissertation. In 1936, while Weil was in America, Chevalley taught his classes for him at the University of Strasbourg. In 1938, Chevalley traveled to Princeton to spend the year at the Institute for Advanced Study, but as war was imminent, the French ambassador to the United States counseled him to remain in the United States.

JEAN DELSARTE WAS born on October 19, 1903, in Fourmies, a small village in the north of France. He received a scholarship that allowed him to study at a good high school in Rouen, where he excelled in mathematics and was considered a brilliant student. He was then accepted to study mathematics at the École Normale Superieure in Paris and it was here that he met André Weil. Delsarte graduated in 1925 and began his doctoral work. When his thesis was accepted, he took a teaching position at the University of Nancy, where he remained throughout his entire career. Before doing so, he discharged his military obligation as an officer in the French army. He kept close connections with his two friends at the

University of Strasbourg, André Weil and Henri Cartan. Delsarte played a major role in the "war of the medals," a campaign by mathematicians to do away with a proposal to award scientists and mathematicians medals of merit. He also organized the eastern section of the French Mathematical Society, in Nancy.

In September 1939, Jean Delsarte was called to duty to defend his country in the Second World War. A captain, he led a battery and valiantly managed to prevent major loss of life against the far better equipped and trained German army he faced. He led a retreat of his force from Alsace to Nîmes, where his unit was dismantled. He spent 1940 and 1941 teaching at the University of Grenoble, standing in for a professor of mathematics named Favard, who had been taken prisoner of war. Clandestinely, he made his way through wartime France all the way to Nancy, where he resumed his work as a mathematics professor. His work was in the theory of numbers, as well as in function theory and problems in mathematical physics.

JEAN DIEUDONNÉ WAS born in Lille, in northern France, on July 1, 1906. When the Germans occupied the city in 1916, during the First World War, Dieudonné and his mother and sister were evacuated and took refuge in Switzerland. His father was then in service, fighting the Germans, and when he was released from the army, the family moved to Paris, where they remained. The boy attended high school in the French capital, but his father wanted him to learn English, so he sent him to Britain, to a school on the Isle of Wight, where he spent the academic year 1919–20. It was here that Jean Dieudonné was smitten with the charms of mathematics. He returned to

France the following year, going back with his family to Lille. He graduated from high school in his hometown, receiving a high honor prize in mathematics. In 1924, he started his studies of mathematics at the École Normale Supérieure, where he met and befriended André Weil. Dieudonné graduated in 1927, passing his examinations brilliantly and taking first place among his classmates.

Dieudonné did his military service as an officer, and after completing it he won a scholarship to Princeton University. He spent a year there, and returned to continue his studies at the École Normale. He then was awarded a Rockefeller Foundation grant allowing him to study for several months at the University of Berlin and later at the University of Zurich, where he worked with George Pólya. Dieudonné received his doctorate in 1931, with a dissertation on problems of polynomials and functions of complex variables. He then took up a position at the University of Rennes, where he remained until some time before the outbreak of the Second World War.

ALL OF THESE mathematicians were very closely associated with the work of Nicolas Bourbaki. But the war disrupted their work. One of them was in safety in the United States, and would return only after the war; some of them were fighting for France and freedom; some of them were hiding from the police; and some of them were continuing their academic lives despite the difficulties.

During the grim period in which Alexandre Grothendieck was growing up, suffering from hunger and other privations in the camps, he would spend all his time studying mathematics. One day, a mathematics tutor came to the village to help the few students there. His name was Laurent Schwartz. Many years later, Schwartz would become Grothendieck's dissertation advisor and mentor at the University of Nancy. The two of them would become very close friends and through Laurent Schwartz, Alexandre Grothendieck would become acquainted with the work of Bourbaki. But history would prove that no one would contribute to Bourbaki's ideas more profoundly that Grothendieck.

ANDRÉ WEIL MANAGED to see his wife on his way from prison to an infantry unit in Cherbourg. She arranged to stay at a hotel nearby, and when the prisoner was set free he joined her there. The next day, his parents and sister joined them, and they all spent the day walking along the Seine. Then André was off to Cherbourg, where he began the monotonous and drab life of an infantry private in an army that had no chance at all of winning a battle.

His unit was stationed in the countryside, and they did very little. Every day or two enemy planes would appear in the sky on reconnaissance missions. Weil was a poor soldier, and was often punished by his commander, a dry and unpleasant man who had no patience with anyone who did not follow orders precisely. Soon Cherbourg was evacuated as the Germans made their advance on the city. In a new position by a small village, Weil was responsible for a machine gun, but his days were spent doing almost nothing. He took baths in a creek, studied mathematical papers, and slept a lot during the day.

Soon the French army lost even these last remaining positions it had held in the countryside, and the soldiers were summarily evacuated by boat to Britain. Weil and his comrades were sent to a camp in central England, near Stoke-on-Trent, where many French soldiers fleeing the Nazi forces were sent.

They spent their days in camp doing very little, and always waiting for news about the fighting, hoping to return to their homeland. Weil, who just before the war began had hoped to find a way to escape to the United States or to find refuge in a neutral country, now badly wanted to return home.

Soon an opportunity to do so arose, as an agreement was reached between the Allies and the Nazis to allow a hospital ship to sail from England to France. André Weil managed to arrange passage aboard this ship by pretending to be sick with pneumonia. This entailed having X-rays taken, but soon he was on his way back to France. He was given comfortable quarters on the hospital ship *Canada,* sailing for Oran, on the Algerian coast, and then continuing to Marseille. As the ship docked in Marseille, André's parents were there to meet him. In England, he had been issued false papers showing he was sick, and, therefore, a day after he came ashore in Marseille he was discharged from the army. He was now in Vichy France, not a safe place for a Jew. He again began trying to find a way to immigrate to the United States.

Weil had to resort to clandestine activity to try to bring his wife across the border from the occupied zone of France to the "free" zone of Vichy. Anyone caught crossing the border between the two "Frances" would either be shot or sent to a concentration camp.[4] And communications across the border between the two regions was very difficult. Through his friends, Weil finally made contact with his wife and made arrangements for her to come to the border. A guide was hired—someone who made a living smuggling people across the border between the two zones. At the appropriate moment, a young woman distracted the attention of the SS guard. Immediately the guide, followed by Eveline, her mother, and her son, slipped across the line into Vichy and the couple was finally reunited. Then Weil, who had no papers, applied for a Vichy passport, which, he hoped, would allow him to travel with his wife and stepson to the United States.

Weil contacted all his American friends, trying to get an offer for a position in the United States. He obtained an invitation to the New School for Social Research, and was told to apply for a non-quota visa at the nearest U.S. consulate. The visa application was denied by the consular officer in Lyon because a non-quota visa would have required the applicant to be the holder of a position in France, and Weil had earlier been dismissed from his professorship at Strasbourg. His application for a Vichy passport had been approved, so technically he could travel, but he still needed a visa. In the meantime, his parents had booked passage to the Caribbean island of Martinique for André, Eveline, and her son Alain.

Weil contacted the New School in New York, and was told to sail right away for Martinique and to try to obtain the necessary U.S. visa upon arrival there.[5] The three of them left Vichy for Marseille. The city was full of refugees at that time, all of them hoping to obtain visas for the United States. Fighting the crowds, Weil managed to find his way into the American consulate. Against regulations, the consul issued him a quota visa to the United States, and André, Eveline, and Alain prepared to embark.

Sailing under the yellow flag of the Armistice Commission, the *Winnipeg* left Marseille harbor, passed through the Strait of Gibraltar, and put into port in Casablanca, where it remained for three days. Two weeks later, the ship reached Fort-de-France, Martinique. Weil and his wife and stepson then made their way to the United States. Weil would hold positions in America for the rest of his life, and in 1976 he would retire from a professorship at the Institute for Advanced Study at Princeton. Thus a mathematician strongly associated with Nicolas Bourbaki would forever remain exiled from France.

Who was Nicolas Bourbaki? And why is his work still important in mathematics today?

four

ARRIVAL IN PARIS

WHEN THE WAR in Europe was finally over in May 1945, seventeen-year-old Alexandre Grothendieck was reunited with his mother. He was among the fortunate children separated from their parents during the war—those who both survived the war and had a parent who did so as well.

Hanka and Alexandre moved to the village of Maisargues, in a wine-growing region in southern France near the city of Montpellier. Alexandre enrolled at the University of Montpellier, which at that time was one of the poorest universities in France, to study mathematics. He spent much of his time making up the material he should have learned in high school, but also deriving on his own the elements of measure theory—a mathematical discipline which, unbeknownst to him, had been established around the turn of the century by the French mathematician Henri Lebesgue.

A professor who taught Grothendieck calculus, Monsieur Soula, whom he had asked about discovery in mathematics, told him that "the last open problems in mathematics had all been resolved twenty or thirty years ago by a person named Lebesgue."[1] Grothendieck noted the irony of this statement: Ignorant of this work, he had by himself derived Lebesgue's theory. Of course, Lebesgue's theory was not at all the "last word" in mathematics. And the immense progress made in mathematics during the following decades of the twentieth century—in which Grothendieck would play a major role—is witness to the absurdity of the teacher's statement. Fortunately, the student was undeterred.

While Grothendieck was a student at Montpellier, he and his mother survived on his student scholarship. They both also worked seasonally as day laborers in harvesting grapes at local vineyards, and when the harvest season was over, they worked in the making of the wine. From 1945 until 1948, mother and son lived in the small hamlet of Mairargues, virtually hidden among the vineyards, a dozen kilometers from Montpellier. They had a marvelous small garden: they never had to work at gardening and yet the earth here was so fertile, and the rains so abundant, that the garden produced a plentiful harvest of figs, spinach, and tomatoes. Their garden was at the edge of a field of splendid poppies. Grothendieck remembers his time there with his mother as "la belle vie."[2] Often, however, his mother was sick. Her health was damaged irreparably from the privations she had suffered in the French concentration camps of the war. As a survivor, however, she was entitled to free medical care. Grothendieck and his mother never had to pay for a doctor's services.

Grothendieck found the classes at the university boring and uninspiring. He understood mathematics at a much higher level than did his professors. At the beginning of each semester, Alexandre would purchase his textbooks and immediately read them from cover to cover, assimilating all the information. For the rest of the semester, he hardly ever showed up for class. He had no need to do so.

Grothendieck graduated from the university in 1948. He

had applied for a scholarship to go to Paris, which required a twenty-minute interview with a government official with knowledge of the discipline to be studied. His short interview turned out to be over two hours long, since the official, who knew mathematics well, was amazed at the depth of understanding, analysis, and creativity shown by the student. He wholeheartedly recommended that the State provide him with a scholarship. One of Alexandre's professors at Montpellier was so impressed with his abilities that he gave him a recommendation to one of the greatest French mathematicians at the time, Henri Cartan, in Paris.

At nineteen, Alexandre Grothendieck descended on the mathematical community of Paris like a storm. His formal preparation in mathematics was lacking, to say the least. His school classes in mathematics were inadequate and without continuity, and his university studies from Montpellier were insufficient for understanding the material at the graduate courses and seminars at the École Normale Supérieure in Paris. In particular, Henri Cartan's seminar at this elite school was exceptionally advanced and required a great deal of mathematical knowledge, which Grothendieck did not have. But the ambitious and immensely talented student worked so hard that he was able to assimilate the advanced material despite lacking the prerequisites. At the seminar, he distinguished himself by carrying on conversations with the renowned professor from the back of the room, speaking with him "as if they were equals, rather than a student speaking with a professor."[3]

In his memoirs, Grothendieck wrote:

> When I finally made contact with the mathematical world in Paris, one or two years later, I finally came to

> understand, among other things, that the work I had done in my own corner [at Montpellier] with the means at my disposal, was that which was well known by "all the world" under the name of "Lebesgue's theory of measure and integration." In the eyes of two or three older students to whom I had spoken about this work (and shown my manuscript) it seemed as if I had simply wasted my time, redoing an "already known." But I do not remember being at all disappointed. At that time, the idea of receiving credit or gaining approval in the eyes of someone for work I had done would have been strange for me.[4]

But Grothendieck also realized that in the academic backwaters where he had studied before coming to Paris he had not wasted his time. He learned on his own things that would prove important in his life: he learned that he was a mathematician. Most importantly, Grothendieck reflected, "I learned, in those crucial years to *be alone*" [his emphasis].[5]

In Paris, Grothendieck fell in with a clique of mathematicians who knew each other extremely well. All of them were very talented, and they worked together creating modern mathematics. But Grothendieck was a lone wolf—he did not work well in a group. He wanted to verify for himself the mathematical facts he came across, rather than accept ideas that people took to be true "by consensus." He felt very welcome, however, within the group in Paris. These were gifted thinkers, and for them mathematics seemed to come effortlessly, while Grothendieck felt like a mole slowly digging its way to a mountain.[6] In hindsight, he realized that "the most brilliant individuals in this group have become competent and well-known mathematicians. However, after the passing of thirty or thirty-five years, I see that they have not left on the mathematics of our time a truly profound imprint."[7]

Grothendieck's work, however, certainly changed modern mathematics. His view is that these mathematicians have not achieved the deep results they might have been able to reach had they not lost the capacity to be alone, to work alone, and to think alone without accepting the word of any authority other than that of their own intelligence.

The group of mathematicians in which Grothendieck found himself, and about which he wrote, included Henri Cartan, Claude Chevalley, André Weil, Jean-Pierre Serre, and Laurent Schwartz. All of them accepted him with friendliness, "without worry or secret disapproval—except, perhaps, André Weil."[8]

This is a telling statement by Grothendieck, for Weil and Grothendieck were certainly opposites. One was privileged, spoiled, selfish, and somewhat lazy—Weil wanted to do only mathematics that was easy for him, and would not set his sights on overly difficult problems. He wanted to enjoy himself, to travel, to socialize, and to be with friends. Grothendieck had a deprived childhood, and everything that ever came his way did so through hard work. He was immensely ambitious, looking for the hard problem rather than the easy way out; and he was not very interested in group work. Another reason why the two men may not have gotten along very well was jealousy. Weil was the best mathematician in the group—until the arrival of the young Grothendieck. Weil must have sensed that Grothendieck could—and in fact, would—go much further in mathematics than he himself ever would be able to. He was suspicious and envious of the new arrival and wanted to keep him at arm's length from his group.

The rest of them supported the young mathematician embarking on an adventure of a lifetime, seeking new knowledge in mathematics. But Weil had always been looked up to by everyone around him. This group of mathematicians in Paris considered him their leader and the most gifted of them all. But the young Grothendieck was on a very different level. Alexandre Grothendieck was very different from Weil in the way he approached mathematics: Grothendieck was not just a mathematician who could understand the discipline and prove important results—he was a man who could *create* mathematics. And he did it alone.

Within a few years, Grothendieck would become a regular member of this group working in Paris. And he would work with other people, listen to what they would say, contribute to discussions, and help everyone around him. But mathematical research he would do all alone. And eventually, he would come to live all alone in this world.

Grothendieck developed an interest in topological groups, an advanced area of mathematics that was being explored at that time, and pursued his own research on this topic. One day he attended a seminar at the university, organized by Nicolas Bourbaki, at which an important French mathematician, Charles Ehresmann, who was an expert in this field, was giving a talk. Grothendieck came up to him abruptly, and demanded, "Are you an expert on topological groups?" Ehresmann, being modest, replied, "Yes, I know something about topological groups. . . ." Grothendieck, turning to walk away, said, "But I need a *real* expert in this field."[9]

MATHEMATICS WAS NOT the only field seeing a spurt of growth and revival in Paris right after the end of World War II. The city teemed with cultural activity, as if the war had stopped all of it, and now people could not get enough. The new freedom from tyranny and persecution brought out French intellectuals in droves. To these were added expatriate philosophers, writers, and artists from around the world, and they brought Paris back to its old glory of the prewar years, a time during which artists, writers, scholars, and intellectuals had made it the capital of culture. Intellectual life was blooming in the cafes of the French capital and on its streets, and the undisputed leader of the Paris café intellectuals was the philosopher Jean-Paul Sartre.

Reviving French thinking from the ashes, Sartre brought philosophy to the street. Existentialism became the reigning philosophy. Sartre, together with Simone de Beauvoir, founded a journal, *Les Temps Modernes*, in which he and Beauvoir and others reviewed important cultural works. He also founded the newspaper *Liberation*, which still exists and thrives in today's France.Yet, his success made him vulnerable to attack, and during the late forties and early 1950s, Sartre had a number of ruptures with former allies, which left him and his new philosophy even more vulnerable. In 1949, Beauvoir reviewed in *Temps Modernes* a book by a young anthropologist, Claude Lévi-Strauss. Her review was very positive, but as history would show, the philosophy engendered by Lévi-Strauss would bring an end to existentialism, and inaugurate a new philosophy in its place: *structuralism*.

This new approach to life, the sciences, and the human condition would begin as a principle in scientific investigation, but within a few years it would sweep all areas of intellectual life. At its kernel, the idea of structuralism would be a mathematical one, but it would slowly take over the sciences, the humanities, economics, and philosophy. Structuralism would begin from strict mathematical considerations, but would widen like a tidal wave—sweeping everything in the intellectual world before it. The ideas of the new philosophy of structuralism would be seeded in the very seminars in the French capital that Alexandre Grothendieck was attending in the late 1940s. Its creators would be virtually all the professors he was meeting in Paris: Cartan, Weil, Chevalley, and others. Café life in Paris would, therefore, be dominated not only by the philosophers and the writers and artists, but also by the mathematicians. And perhaps for the first time in modern history, mathematics would play a key role in the general culture—in a way that it did only in the very distant past of ancient Greece.

Since Sartre and his allies were decidedly nonmathematical in their approach to life, they would inevitably be left behind. Their philosophical theory of existentialism would end its reign as the strictly axiomatic, rigorous, and system-oriented theory called structuralism swept France and the rest of the Western world. The mathematicians would play a key role in the new milieu not only as proponents of a new and widely used approach to life, but also as *connectors* among practitioners in different fields: the exact sciences, the social sciences, art, literature, psychology, economics, and philosophy. This would launch a new age for mathematics, one in which the role of the discipline in our culture could not be matched by any other. For, while the mathematicians strove to keep their discipline abstract and abstruse—appearing to be completely disconnected from the real world and oblivious to everything that was taking place around them—in fact, the ideas developed by these mathematicians sitting in Parisian cafes would prove to be of crucial importance for society as a whole. The ideas of these mathematicians would constitute nothing less than a revolution in human thought—one whose effects would be felt far and wide. When this revolution reached its culmination in the late 1960s, not a single area of human interest would be left untouched.

DURING THIS PERIOD, in Paris, Grothendieck was completely oblivious to the changes about to take place around him. And yet, his work and his ideas would, within a few years, form the vanguard of the new philosophy. But Grothendieck's mind was so supple and so fecund that he would forever remain far ahead of his own time. Unbeknownst to him during this early period, he would within a few years find the defects in the new theory that would sweep Western thought. He would single-handedly try to correct the deficiencies of the mathematical theory forming the backbone of the new philosophical and scientific approach, structuralism.

And yet, the mathematicians now philosophizing in Paris cafes alongside writers and intellectuals would find it impossible to follow Grothendieck's lead. Frustrated with the lack of direction, he would abandon the group and set out on his own. But he would be unable to help the humanities, philosophy, social sciences, and economics recover from the conflict. Structuralism would, therefore, decline just as existentialism did, and both would give way to the new trend in Western culture: postmodernism. But this was still decades in the future. In the meantime, there was very exciting work to do—work that would affect every discipline in our world. The end of the war truly spelled a new beginning for Western culture.

DESPITE HIS GENIUS, Grothendieck sorely needed a better background in mathematics. Henri Cartan, who recognized this fact, suggested that Alexandre leave Paris and that the young genius move to Nancy to work toward his doctorate at the University of Nancy under the extremely gifted and renowned professor there, Laurent Schwartz. Grothendieck followed Cartan's suggestion and moved to Nancy. There, he distinguished himself in his work as a graduate student, greatly impressing Schwartz and the other professors at this school. Grothendieck produced *six* seminal papers of mathematical scholarship, each of which would have been considered a masterpiece and would easily have qualified as a doctoral dissertation. Although his professors found it difficult to choose which of these should be used as his dissertation, he was awarded his doctorate in mathematics from the University of Nancy in 1953.

At Nancy, he was a frequent guest of Laurent and Marie-Hélène Schwartz and felt at home in their house. Schwartz and Dieudonné gave Grothendieck a number of problems in topological vector spaces, whose theory was being developed at the time, and they were astonished to find out within a few months that the young student had solved each of these exceptionally difficult problems; he thus advanced this new field significantly.[10]

Grothendieck lived in Nancy with his mother, who was suffering from tuberculosis, a disease she had contracted in the concentration camps. The mother and son were renting rooms from a woman who was older than the son, and Grothendieck developed a relationship with her. Through this liaison, his first son, Serge, was born.

After receiving his doctorate, Grothendieck returned to Paris and again joined the seminars at the École Normale Supérieure as well as at the Sorbonne. In addition to Henri

Cartan, who ran this important seminar, Grothendieck came to know better all the mathematicians in the group associated with the work of Nicolas Bourbaki. But just who was Nicolas Bourbaki?

five

THE GENERAL AND THE ZEITGEIST

FRENCH GENERAL CHARLES Denis Sauter Bourbaki was born in Pau on April 22, 1816, and died on September 27, 1897. He came from a prominent Greek family, some of whose members had immigrated to France in the late eighteenth century. His father was a Greek colonel who died in the Greek war of independence in 1827. As a young man, Charles Bourbaki was educated at the military academy, the École Spéciale Militaire, and upon graduation took part in the African Campaign of 1836 to 1851 as a lieutenant.

Later, in 1854 to 1856, Charles Bourbaki served France in the Crimean War, commanding some of the Algerian troops in that war, and performed his duties so well that in the course of the war he was promoted to the rank of Brigadier General. He took part in the battles of Alma, Inkerman, and Sevastopol. It was here that the Bourbaki name became famous due to the general's bravery and perfect mastery of the art of war.

Bourbaki went on to serve in Algeria in 1857, and following that campaign was promoted again, this time becoming a division general. He then took command of Lyons in 1859.

In 1860, General Bourbaki commanded French forces

General Charles Bourbaki
(POUR LA SCIENCE)

in Italy, and continued his spiraling ascent up the military hierarchy to become inspector general of all the French infantry forces. In 1862, General Bourbaki was offered the throne of Greece, but declined. In 1870, Emperor Napoleon III made Bourbaki the commander of the Imperial Guard. Soon, he was commanding the battle of Metz.

The French were defending the city of Metz from the besieging Prussians. The Prussians tricked Bourbaki into traveling to England, causing him to believe that a peace conference was being arranged there. Once he arrived in Britain and met with Empress Eugénie, who had taken refuge there, the general realized he had been tricked. He immediately returned to Metz. As the French were quickly losing the war, Bourbaki was transferred to command the Army of the East, a ragtag army that was poorly equipped and trained. With these troops, General Bourbaki tried to lift the siege of Belfort. But his forces were repelled by the Prussians, and further German forces joined the chase after the general and his army. Bourbaki's army was now driven toward the Swiss border. The remnants of his troops were starving, and, of 150,000 men, only 84,000 remained alive.

The general led the retreating French army to Besançon. In a valiant attempt to escape the pursuing Prussians, Bourbaki took his forces across the border into Switzerland; but there they were captured and disarmed by the Swiss. On January 26, 1871, Bourbaki relegated his command to General Clinchant and attempted suicide—but failed. The pistol bullet he had fired into his own skull was flattened on impact and deflected, and he survived the attempt. Clinchant took him to receive medical help in Switzerland, and after the humiliating French surrender to the Prussians, General Bourbaki returned to France.

Bourbaki was then given one last command: In July 1871 he again took command of the city of Lyons. Some time later he was made the military governor of Lyons.[1] In 1881 he was placed on a list of officers that were to be retired from service. In 1885, General Bourbaki tried unsuccessfully to run for a seat in the French Senate.

FRANCE REMEMBERED ITS valiant general from an immigrant family, and his likeness was preserved in portraits and statues at various locations in France. In the town of Pau, for example, there is a street named rue Bourbaki. And it was thus that Raoul Husson, a history buff who was also a third-year mathematics student at the École Normale Supérieure in Paris, learned about the general.

In 1923, Husson prepared his trick. Every year, the third-year math students at the ENS would devise a mathematical prank with which to taunt the first-year students: the unsuspecting fresh recruits to this elite school. Husson walked into the classroom full of math freshmen dressed in a uniform and wearing a long false beard. He wrote on the board:

> Theorem of Bourbaki:
> You are to prove the following. . . .

None of the students could even begin to understand, let alone prove, the "Theorem."

DURING THAT SAME period, French students used to hang around the wide boulevard that marks the southern boundary of the center of Paris. In this area, there was a large mound of garbage that grew bigger by the day. The students named it Mount Parnassus, after the mountain in central Greece dedicated to Apollo, at the foot of which lies Delphi. Eventually, the boulevard became known as the boulevard of Mount Parnassus, or simply (in French) the Boulevard Montparnasse. This area became the gathering place of students and intellectuals and a favorite place for playing tricks on unsuspecting passersby.

One day in 1923, an announcement was made throughout Paris that the prime minister of Poldevia would be making a public address at the Boulevard Montparnasse. A large crowd gathered, and a student purporting to introduce the "Prime Minister of Poldevia" gave a speech in which he bewailed the abject poverty of the nation of Poldevia, exhorting his listeners to donate as much money as they could spare for the welfare of a nation in need. Hats were passed around, and the Parisians gave money. "And now," the student announced, "I present to you the prime minister of this unfortunate nation, whose citizens are so poor that they cannot even afford a pair of trousers." The "Prime Minister," who then climbed onto the makeshift stage, was seen by all to be in his underwear.

André Weil was one of the students at the ENS that fall day in 1923 when Raoul Husson presented his "Theorem of Bourbaki." Weil was very much impressed by Husson's prank. He also knew what happened on the Boulevard Montparnasse when his fellow students played their vulgar trick on passersby, and the name Poldevia remained etched in his memory.

In 1930, Weil became friends with the young Indian mathematician D. Kosambi, fresh out of Harvard, whom he had met in Benares and later hired for a position at the Aligarh Muslim University at which he worked. Weil told his new friend about the "Theorem of Bourbaki" and about the "Nation of Poldevia" and suggested that Kosambi write a mathematical article, the contents of which would be the review of imaginary works of a mathematician named Bourbaki, a member of the Academy of Sciences of Poldevia.

Kosambi proceeded to write an article entitled "On a Generalization of the Second Theorem of Bourbaki," a completely fictitious work of mathematics, which he managed to get published in the *Bulletin of the Academy of Sciences of the Provinces of Agra and Oudh Allahabad.* Kosambi attributed the "theorem" to "the little-known Russian mathematician, D. Bourbaki, who had been poisoned during the Revolution."[2]

This seems to be the first mathematical publication—albeit nonsensical—ever attributed to Bourbaki. Eventually, *D.* would be replaced by "Nicholas," and Russia would be replaced by the imaginary country of Poldevia as Bourbaki made his grand appearance on the world stage.

BOURBAKI AND HIS contributions were a direct product of the cultural upheaval that took place in the early decades of the twentieth century. Pure mathematics might seem like an abstract field of human study with no direct connection with the real world. But, in reality, mathematics is closely intertwined with the general culture. Few mathematicians, even those doing the most abstract and obscure work, are completely divorced from the realities of the general culture around them. So major developments in mathematics have often followed important trends in popular culture; and, conversely, developments in mathematics have acted as harbingers of changes in the general culture.

A REVOLUTION TOOK place early in the twentieth century in the field of physics, and it changed the way we view the universe. This transformation, which would encompass all aspects of life, began in 1905, when Albert Einstein changed science forever by stripping the old ways of viewing the universe, thus creating the special theory of relativity. He would follow this achievement ten years later with his comprehensive theory of gravity, general relativity. Einstein deconstructed physics by finding a completely new way of looking at physical space and time and their properties, redefining structure in physics by recognizing the three elements:

Space
Time
The speed of light.

Einstein's genius understood that the speed of light was constant, while both space and time were malleable and could be warped and wrapped around the universal constant, the speed of light.

Thus, Einstein was able to throw away the old science, reconstruct our understanding of the universe from first principles, and then build everything up from this new foundation, achieving results of great complexity based on this principle. We would never view the universe the same way again after the great leap afforded us by Einstein's theories of relativity. No longer would something as natural and as intuitive as time be taken as unchangeable: time may be different for two different people, depending on their speed or on the force of gravity they feel. Thus, Einstein's "twin paradox," which proves that a twin flying in a speeding spaceship ages more slowly than the twin who stays on the ground, forever shattered our old understanding of time. It opened the door to completely new ways of thinking about space, time, and the universe.

The twentieth century started with a discarding of the old ways: both for good and for bad. The century saw two world wars, which spelled the end of the world of innocence we knew until then. Science was changed by Einstein, and art and humanities and philosophy and economics would all follow this major break with the past.

While the world was still trying to adjust its ways of thinking to Einstein's new theories, another revolution took place in physics. In 1926, following ideas first put forward by German physicist Max Planck in the year 1900, an international group of very young physicists (mostly in their twenties) working both together and independently, brought us the quantum revolution. While Einstein had showed us that the very large and the very fast do not obey the old laws of Newtonian physics—worked out two centuries before Einstein's time—the quantum physicists Bohr, Heisenberg, Schrödinger, Pauli, and Dirac showed us that on a micro-scale, the world is not as we think it is, either.[3]

The laws of Newtonian physics break down as we reach the small scale of atoms and electrons and protons. In this world, what is here is not necessarily here, and what is now may not really be now: place and time are malleable, and everything is seen through a haze of probabilities, rather than with certainty.

Causality no longer works in this world either. In the quantum realm, it is not always possible to determine that action A caused result B. The breakdown of this principle, one that people rely on in everyday life and science and philosophy, is the greatest shock to the way people think. For once we lose causality, the world around us really makes no sense at all.

The twin revolutions in twentieth-century physics, whose seeds began in the first decade of the new century, made people realize that we cannot rely on the past, on the way human thinking had been done for centuries, and that a new era must begin—one in which a complete break with the past is called for.

In 1900, the German mathematician David Hilbert—one of the greatest mathematicians of his day—gave a keynote lecture at the Congress of Mathematicians that was convened that year. Hilbert exposed his view as to the future of mathematics in the twentieth century.

In particular, Hilbert presented ten problems, now called "Hilbert's Ten Problems," which he thought should be solved in the new century. Hilbert's program was a systematic development of mathematical ideas in the new century, and as part of the new developments in this field, the ten key problems—difficult problems that had resisted solution for some time—should be solvable, Hilbert believed. Hilbert's program augured a revolution in mathematics. For in order to solve the ten problems and others (the list had grown some more), mathematics had to make great leaps forward. The state of mathematics at the turn of the twentieth century was not good.

There were good mathematicians at various places: Poincaré, Lebesgue, and Hadamard in France; and there was a good German group, which included Hilbert and Emmy Noether in Göttingen, as well as Weierstrass, Dedekind, and Kronecker in Berlin, and Cantor in Halle. But there was no program to revamp mathematics and bring it into the new century and in line with new developments in other fields. Hilbert's vision was to make mathematics more precise and rigorous, so that the problems he outlined in his talk could indeed be solved. Mathematics was ready for a new beginning.

German mathematicians would have been ready for the change, except that their groups were at separate locations, and they did not agree with one another. Internal fighting dominated German mathematics, making it difficult for new ideas to get accepted. In particular, Georg Cantor, a genius who single-handedly developed modern set theory, found it impossible to get his ideas accepted by the other German mathematicians. His ideas were so advanced, and so contrary to human intuition, that the others, in particular Leopold Kronecker, persecuted him for these views of mathematics.[4]

In France, this change would also be difficult to bring about because French mathematics had been dominated by the intuitive, rather than precise, approach of Henri Poincaré. Poincaré was a mathematician of deep insight and great ability. He was called "the last universalist" because he had mastery of so many areas within mathematics. But he stressed intuition and approximation rather than a precise, rigorous approach that could lead to good results in terms of correct proofs of theorems and a firm foundational basis for mathematics.

Clearly, however, a new wind was blowing at the beginning of the twentieth century. The old ideas were being discarded daily and new approaches were being adopted. In the same year of the turn of the twentieth century, 1900, another event that would turn out to be of great importance in the history of science and ideas took place. That year, Ferdinand de Saussure began to establish the foundation of the new science of linguistics. As the century moved on, more work would be done in linguistics, building on Saussure's foundation. And eventually, within a few decades, the seemingly unrelated work begun that fateful year in physics, mathematics, and linguistics would be joined to work in other fields, all leading to a new approach to human thinking.

six

WHEN BRAQUE MET PICASSO

TWO YEARS AFTER Albert Einstein first brought us his theory of special relativity and forever shattered the way we view the universe, Pablo Picasso and Georges Braque did the same for art. They threw away the old ways of viewing art, and deconstructed art as a new discipline, thus creating modern art.

Cubism and other forms of modern art went back to first principles of shape and form and denuded a subject of its unnecessary details, leaving the very basic elements of a figure. Like Einstein, they concentrated on the *relationships* among the elements of a subject, as space and time related to the speed of light in relativity theory, creating a new art form.

There were mathematical and philosophical ideas at play in both the new physics and the new art. Braque's *Femme à la guitare*, 1913 (Centre Georges Pompidou, Paris), demonstrates this break with the past in painting and the emphasis on the simplest visual elements and their structure and the interrelations among them.

But cubism was born five years before this painting was made, when in 1907 Picasso created *Les Demoiselles d'Avignon*, the first cubist painting. This picture broke the mold and exhibited the beginning of the destruction of the old and

Georges Braque's *Femme à la guitare*
(CENTRE GEORGES POMPIDOU, PARIS)

the restructuring of the new to begin a new age in painting, one that would eventually strip away form and redefine structure.

Picasso's African period, 1907–1909, saw him go back to primitive tribal art to seek the new elements of space he wanted to introduce into art after discarding the old realist way of painting. This was one stage in the development of the world's most unusual artist, and it would lead to the invention of cubism by Picasso and Braque.

Pablo Picasso was born in Malaga, Spain, in 1881, the son of an art teacher. From an early age, he showed a great ability

Picasso's *Les Demoiselles d'Avignon*
(Museum of Modern Art, New York)

and interest in drawing. He was exposed to traditional forms of art, but as he grew older, he embraced the "modernista" movement in Catalonia. Picasso studied in art academies in Corunna, Barcelona, and Madrid, and became disaffected with academic art.[1] He moved to Barcelona as a young man and pursued the study of painting that was influenced by the French avant-garde movement. These young artists emulated the styles depicted in the art of Toulouse-Lautrec, and pursued the life of anarchists and bohemians.

In 1900, nineteen-year-old Picasso visited Paris, the city he had longed for as a teenager, where he visited the

Universal Exposition, and was charmed beyond words by the French capital. After his return to Barcelona, he vowed that he would find a way to move to Paris. He realized this dream the following year, when he came back to Paris to work on his first major exhibition, arranged by Ambroise Vollard, who had set up Cézanne's important exhibition of 1898. Picasso's exhibition was fabulously successful, although some accused the young Spanish painter of imitating known artists such as Toulouse-Lautrec. In part because of this criticism, Picasso changed tack. He dropped his vivid colors and instead chose a somber blue palette, which inaugurated his now-famous "Blue Period."[2] This would later be transformed into the browns, tans, and grays of the monochrome phase of cubism.

Picasso began selling his paintings with much success. Drawn to dramatic and symbolic topics for his paintings, in his "blue period" he created major works that depicted love, desire, birth, and death.

Georges Braque was born on May 13, 1882, in Argenteuil-sur-Seine, France. He grew up in Le Havre and studied at the local École des Beaux-Arts until 1899. Then he left for Paris to study under a master decorator, and received his craftsman certificate in 1901. From 1902 to 1904, Braque worked as a painter at the Académie Humbert in Paris, and by 1906, his work, which had been associated with impressionism, was no longer in that tradition and he adopted the Fauve style. Braque spent the summer of 1906 in Antwerp with Othon Friesz, and when he returned to Paris at the end of the summer, he exhibited his Fauve paintings at the Salon des Indépendants.

Braque was interested in painting techniques and pictorial order that were studied by Cézanne. Picasso, however, was not interested in Cézanne but was rather influenced by the exotic imagery of Gauguin, whom he emulated in his paintings during this period. Picasso settled in Paris in 1904 and became the center of a social circle named after him, "La Bande á Picasso." Braque and Picasso inevitably became interlinked through their relationship to Matisse. Braque had almost become an apprentice of Matisse and was greatly influenced by his work, while Picasso saw Matisse as a rival. Picasso and Matisse met at the Salon des Indépendants in 1906 and competed with each other to become "king of the moderns."[3]

The two young artists, Picasso and Braque, lived and worked within the same modernist circle in Paris. Sometime in 1907, Braque left two calling cards for Picasso. He scribbled "regards" on one and "anticipated memories" on the other. Picasso's sketchbook in early 1907 contains the entries: "Write to Braque," followed by, sometime in April 1907: "Braque. Friday." Later that year, the poet Guillaume Apollinaire accompanied Braque to Picasso's studio to see a new and unprecedented painting: five nude women drawn in the proto-cubist style—*Les Demoiselles d'Avignon*.

There followed a period of intense work for them both, during which Braque and Picasso further developed cubism, working as friends and as friendly competitors. From 1909, Picasso and Braque entered a later period in which they developed cubism in a new direction, and by 1911 their styles, up to that time decidedly different, became extremely similar. In 1912 the two artists started to incorporate collage elements into their paintings and to experiment with the *papier collé* (pasted paper) technique. Their collaboration continued until World War I broke out in 1914, when Braque joined the French army. He was wounded in battle, and upon his recovery in 1917 he began a close friendship with Juan Gris. The paths of Braque and Picasso then diverged.

The revolution in art brought about by Braque and Picasso was summed up as follows:

> "Let each person proclaim: here is a great negative work of destruction to be accomplished." This exhortation, made in the "Dada Manifesto, 1918" read by Tristan Tzara on March 23, 1918, in Zurich, gives a good idea of the atmosphere presiding over modernity's beginnings. Destroy the old world in order to reconstruct a new, modern world! Depose the old mankind in order to move towards a new one! Dismantle the Academy in order to institute a new state of mind. Such was the program proposed by modern art, a program that placed the idea of destruction at the very heart of the definition of art. And this redefinition has continued to be effective throughout the century.[4]

The revolution took place in other aspects of modern culture as well: sculpture, architecture, theater, music. And it would continue to anthropology, linguistics, psychology, literature, and philosophy, and eventually cover all aspects of life in the twentieth century. Nicolas Bourbaki would play a major role in this upheaval, for he would bring these new ideas to mathematics, thus creating a whole new approach to the field. The change began with a destruction of the old and with the invention of a new approach.

> How are we to turn pure negation to affirmation? One way is to come up with a comprehensive formalization. Why? Because pure form, shown as form, and

> pointing to its own rules, does indeed seem to abolish all reference to content as well as any imitative or representative approach. In literature this idea already existed with Flaubert. The twentieth century broadens these vistas and plunges into radical endeavors of formalization-destruction. In the first way, it does so in the constructivist and geometric way of Malevich and Mondrian. This is in the school of the axiomatic and formalist way exercised in the mathematics of the century, Hilbert's program and Bourbaki's monumental treatise.[5]

Einstein's dramatic discovery in 1905 dealt a fatal blow to our view of the universe. His theory added a fourth dimension to our world: the dimension of time. It was now time for our new view of nature to enter into art. The mathematician Maurice Princet explained Einstein's theory to artists, and they began to mimic these discoveries in physics in their own work.

While Princet did not create cubism, his explanations did influence the new art form. Picasso, Braque, Duchamp, Metzinger, Gris, and others created the new art form by destroying the old perspective and realism and creating the new way of painting. But the fourth dimension itself was now to enter art as well. Artists tried both to paint in a way that seemed to reflect a new dimension and to create art in which time itself was the variable. Time was the added dimension. Marcel Duchamp, in particular, explored paintings in which subjects were depicted at different points in time. He even experimented with paintings in which time was accelerating or slowing down.[6] Duchamp's *Portrait de joueurs d'échecs,* 1911

(Philadelphia, The Museum of Art, Arensberg Collection), depicted below, serves as a good illustration of these ideas.

In 1911, Duchamp read Elie Jouffret's book on geometry in four diemsions, *Traité élémentaire de géométrie à quatre dimensions* (1903). Since by then Einstein had made the fourth dimension a real element of our universe, Duchamp decided to explore it in his art. He painted a series of studies and canvasses of checkers and chess players, the painting shown here being the main one.This cubist painting shows his two brothers, Villon on the right and Duchamp-Villon on the left, engrossed in the game. The two figures are fused in the center, creating

Marcel Duchamp's *Portrait de joueurs d'échecs*, 1911.
(Philadelphia, The Museum of Art, Arensberg Collection)

an illusion of a single point of view, thus doing away with the old spatial perspective. The figures reappear, deformed and at changing angles, higher on the canvass, reflecting the effects of time as a fourth dimension. The two players, however, are unaware of the fourth dimension and are locked in a static subspace of the dynamic universe around them. This painting exhibits the typical monochrome tonalities of cubism.[7]

THE TWENTIETH CENTURY began with a total abandonment of the principles that had guided society over many centuries. In physics, the understanding of the universe changed completely and abruptly with the emergence of two theories: Einstein's theory of special and general relativity, and the quantum theory, its roots laid in 1900 by Max Planck. Art changed completely and drastically at the beginning of the twentieth century with the work of Picasso, Braque, Duchamp, and others. It is also well worth mentioning that in the years 1916 to 1923 another movement in art and culture appeared, called Dada. Dadaism was an art form that did away with traditional concepts and attempted to reshape the way the world was perceived.

Architecture was changed as well in the first decades of the twentieth century. In fact, the upheaval affected all areas of human creativity and analysis: the sciences, mathematics, art, architecture, theatre, philosophy, politics, and so on. In politics and economics, the theories of Marx and Engels developed and took root. And in psychology, the pioneering works of Jung, Adler, and, most importantly, Sigmund Freud forever changed the way we think about the human mind, and the way we understand its workings.

All of these abrupt ruptures with the past caused a shakeup

in everything people believed. There was a clear feeling in the air, the Zeitgeist of the period, that everything in the past must be discarded and society must start afresh. With this scientific, philosophical, economic, artistic, and humanistic break with the past, war would inevitably come. But between wars, culture would flourish and rewrite itself completely. Within this milieu, mathematics would play a major role—more than it ever had in human history since the Classical Age of Greece.

seven

THE CAFÉ

IN 1933, ANDRÉ WEIL started teaching mathematics at the University of Strasbourg, where his good friend from the École Normale, Henri Cartan, was also on the mathematics faculty. The books used for mathematical instruction were not adequate, and the curriculum was in bad need of revamping. The two young professors would complain to each other about these problems they had with teaching elementary college mathematics. In particular, as Weil later put it, he became annoyed by the constant questions about how to organize course material put to him by Cartan.

Describing an idea that occurred to him in this context, Weil wrote, "One winter day toward the end of 1934, I came upon a great idea that would put an end to these ceaseless interrogations by my comrade. 'We are five or six friends', I told him some time later, 'who are in charge of the same mathematics curriculum at various universities. Let us all come together and regulate these matters once and for all; and after this, I shall be delivered of these questions.' I was unaware of the fact that Bourbaki was born at that instant."[1]

There was, of course, an entire new generation of young mathematicians, products of the École Normale and other

French universities, who were now taking positions at universities, and Cartan and Weil were only two members of this group. This was the post-World War I generation of French mathematicians.

The Great War had decimated France's intelligentsia and academia. The numbers of French university graduates who died in that war was staggering. About half the graduating class of the years 1910–1916 died during the First World War.[2] There followed a period of stagnation when new mathematical and educational ideas were sorely needed.

Weil had had excellent connections with German mathematicians, forged during his numerous travels in Germany, and he knew that, despite its difficulties, German mathematics was forging forward: new ideas abounded there, and groups of mathematicians were working together at German universities, such as the excellent research groups Weil had visited in Göttingen and Berlin a few years earlier.

There was no reason why something similar should not be possible in France. The stark inadequacy of prewar textbooks and syllabi in all French universities—along with Cartan's ceaseless questions—were the catalysts that gave Weil the idea of organizing all his colleagues and taking common action. Together, they would do away with the past and start afresh.

Weil called a meeting for noon on December 10, 1934, to be held at the *Café Grill-Room A. Capoulade* (which doesn't exist anymore; today it is the site of a fast-food restaurant) at 63 Boulevard Saint-Michel, at the corner of the Boulevard Saint-Michel and rue Soufflot, across from the beautiful Luxembourg Gardens and near the Panthéon, in the heart of the university area in the Latin Quarter of Paris.

In the below-ground level of this café came together the

young mathematicians Henri Cartan, Claude Chevalley, Jean Delsarte, Jean Dieudonné, René de Possel, and André Weil. Among them, they represented the universities of Strasbourg, Nancy, Rennes, and Clermont-Ferrand. All of them happened to be in the capital for a mathematics conference held at the newly opened Institut Henri Poincaré. They embarked on the ambitious project of setting the mathematics curriculum for courses of calculus and mathematical analysis offered in all the universities in France.

This group, collectively, would become Nicolas Bourbaki. So Nicolas Bourbaki was never a single person, even though there had once been a general by the name of Bourbaki. These six young mathematicians meeting in Paris—and others who would join them, while some of them would leave—would in a way continue the pranks of Paris student life by inventing a person and by founding a secret society. They would create for their invented person a family and a family history—hence the visiting cards found in Weil's room in Finland, and the invitation to Bourbaki's daughter's wedding. All of this, including a baptism and a baptismal certificate, godparents, and so on, would be fabricated in order to create a persona—an amalgam of the identities of the individual members of the group. But all this was still in the future.

That December day in 1934 the group lunched on cabbage soup and grilled meats served with *endives braisées* or *pommes soufflées.*[3] André Weil opened the meeting by saying that the goal of the common undertaking was "to define for the next twenty-five years the syllabus for the certificate in differential and integral calculus, by writing collectively a treatise on analysis."

Answering to the fact that such a syllabus may already have existed, he quickly added, "Of course, this treatise will be as modern as possible."[4] As expected, all present were strongly in favor of this common purpose for the group. The mathematicians then discussed the fact that they planned to write such a syllabus *collectively*—without any of it being identifiable as the property of any one of them.

According to Henri Cartan's recollections of this historic meeting, Jean Delsarte was particularly adamant on the issue of the document being the work of a collection of people, arguing that by making it lack any individual elements, the group could protect itself against a possible future claim of intellectual property.[5]

Delsarte also urged the group to produce the document as quickly as possible, and expressed the hope that the book could be put into print within six months. This would assure the element of surprise to the entire intellectual community of France and demonstrate that the "young Turks" of the mathematical community were taking over in a swift and forceful way from the ossified establishment of this profession in France. In the ensuing enthusiastic discussion, someone suggested that the syllabus be a thousand pages long, and that it be geared toward teaching in all the universities of France, rather than constitute simply a work of reference.[6]

Calculus had been taught in France using an old textbook written by Edouard Goursat, which was badly outdated and reflected the stagnant state of French mathematics as a whole. The discipline had not been reinvigorated and was taught and researched as it had been in the 1920s and earlier. No new ideas were being introduced into French mathematics at the time. This was why the meeting at *Capoulade* was so innovative and important. Here was a group of vigorous young men meeting together to rewrite French mathematics. The system was so outdated by this time that professors all over France were allowed to use their own lecture notes, or anything at all they could get their hands on in order to teach calculus and related subjects.

It was, in fact, ironic that the ineptitude of Goursat's calculus textbook allowed mathematicians great freedom in choosing their teaching materials. This group of young men meeting in a Paris café sought to correct this situation and to come up with a good, workable textbook that could be used uniformly to teach calculus and other mathematics courses all over France. It was a very ambitious goal, and it represented the young spontaneously wresting control over mathematics instruction from the older generation.

A discussion ensued over lunch, and Weil suggested that in carrying out their task they not rule out any topic that could possibly be included in the syllabus. The collective decision was then unanimously adopted: The group should not assume anything that had been written in the past: it should literally start from scratch.[7]

At that moment, Jean Delsarte made a suggestion that would have far-reaching implications, not only for the group and for French mathematics as a whole but also for the state of worldwide mathematics in the twentieth century and beyond. Delsarte stated that he believed that the common treatise the group was writing should start in the most *abstract* and *axiomatic* way.

He was thus taking literally the group's decision to start from scratch. Starting from scratch meant starting from the very first principles of mathematics, in its greatest generality and abstraction. While they may not have been aware of it at the time, this small group of young French mathematicians was about to do what Euclid did over two millennia earlier:

write down mathematics starting from the barest set of first principles and build the discipline from this foundation upwards.

The beginnings of mathematics can be seen as the notions of sets and set operations (unions and intersections of sets), so the group decided that its program should begin with these very elementary and intuitive notions that can form the foundation of all of mathematics. Sets and set operations are demonstrated below.

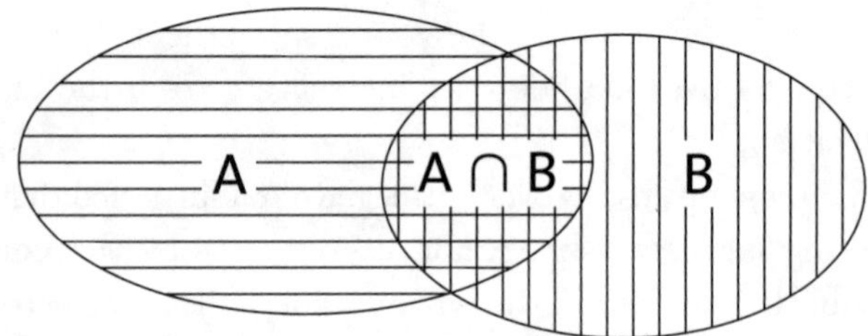

Two sets, their union, and their intersection.

Delsarte's suggestion was enthusiastically taken up by everyone present, and the decision was unanimously taken to start the treatise with an introductory section titled *"Paquet Abstrait"* ("Abstract Package").[8]

But what should follow this abstract package? Each participant suggested his own favorite topic, but no final decision was made as to what to include and where within the entire treatise. As it was getting late, and they all felt that they had already achieved much, Weil closed the meeting, saying that the next meeting should take place a month later, in mid-January, at the same place and the same time. He asked that each participant bring with him to the next meeting a list of topics to be included in the syllabus.

These young mathematicians were faced with interesting problems. Within the course on mathematical analysis, for which Goursat's textbook had been designed, there was the issue of defining the Stokes integral, something that was done very poorly in the text. This led to many other problems, including how to define a curve or a surface, and the idea of *shape*. What is a number? What is a set? These were very important questions that the proposed book had to answer. And there was an important further question: What kinds of applications should be included in the course and the textbook?[9]

The group named itself the Committee for the Analysis Treatise, and the document they planned to produce was called the *Traité d'analyse (Treatise on Analysis)*, analysis being the general term for calculus and related mathematical topics taught at French universities at the time, as well as the term for advanced, theorem- and proof-based calculus taught in American (and other) universities today.

At the next meeting, a few new members joined in: Paul Dubreil, Jean Leray, and Szolem Mandelbrojt (who had immigrated to France from Poland).[10] The committee had now decided on a publisher for its treatise; the members chose a small, independent Latin Quarter publisher who was far from the mainstream and liked to take on new and exciting—if risky—projects. It was the Hermann publishing house, headed by Enrique Freymann.[11]

These young mathematicians were the best that France had to offer. They had all published numerous papers on mathematics, and most of them had traveled to other countries on various academic grants. Weil had traveled to India and to a number of European countries. Others in the group had visited mathematics departments in Germany, Italy, Sweden, Denmark, Hungary, and other countries.

Abroad, they had all made contact with the best mathematicians in the world, and had been influenced by the new ideas prevalent in the world outside France. They had all received prizes, fellowships, and other distinctions in mathematics.[12] Thus, these were the best possible individuals to lead France into the realm of new ideas and to rewrite French mathematics.

So when they arrived for their second meeting, each one of these mathematicians had with him a list of topics he considered important to include in the treatise, topics influenced by the mathematics pursued at host institutions that these mathematicians had visited, and also topics they found to be important through their teaching and research.[13]

According to Claude Chevalley, "The project, at that time, was extremely naïve: the basis for teaching the differential calculus was Goursat's *Traité*, very insufficient on a number of points. The idea was to write another to replace it. This, we thought, would be a matter of one or two years. Five years later, we had still published nothing."[14]

The members of the group were all in their twenties or a little older, and none were well-known at that time. But their little project would soon expand far beyond these early goals, and within a couple of decades the members of the group would become renowned mathematicians whose influence would extend throughout the entire world. Their project would make history and would influence Western culture in ways these young mathematicians could not even imagine.

Paris was the natural place for the committee's meetings, since its members would regularly come to the French capital to attend the seminars at the Institut Henri Poincaré; and the

Latin Quarter, with its demonstrations and political and social chaos, provided a lively backdrop for a group of young mathematicians rewriting French mathematics inside a café.

Their meetings were typically chaotic, with members shouting and interrupting each other but coming to a surprising consensus at the end of each meeting. And so, through several such meetings, the project was progressing. But soon they wanted to expand the scope. The members found that they wanted more—much more.

And, indeed, had this group put together only a mathematics syllabus—ambitious as it may have been—it would not have been remembered by history the way fate would make it remembered. "We must write a treatise that will be useful to all: to researchers (*bona fide* or not), aspirants to posts in public education, physicists, and all technicians. As a criterion, we can say that we should be able to recommend this treatise, or at least its most important sections, to any self-taught student, presumably of average intelligence," they concluded. And they added, "Mostly, we must provide tools, which should be as powerful and universal as possible. Usefulness and convenience should be our guiding principles."[15]

To achieve a set of such disparate goals as setting a syllabus for a course to be taught throughout the land, and at the same time satisfying the needs of practitioners of all sorts—both within mathematics and in other, applied areas—as well as a person "of average intelligence" on the street, the committee had to find an effective way in which to function regularly.

It did so by appointing several subcommittees for tackling different tasks. Each subcommittee numbered from two to three members, and its participants were those people with expertise and interest in the given topic the committee was to study. The organization also appointed officers, and since they would need funds to support the publication, as well as for other expenses, money was collected.

AS SUMMER APPROACHED, the committee decided it needed a change of venue, a change that would also reflect the grand scheme it was now contemplating. A congress was thus called, to be held over an entire week—July 10-17, 1935—in the small village of Besse-en-Chandesse, in the French countryside forty kilometers from the city of Clermont-Ferrand. It was during the time of this congress that Nicolas Bourbaki was born.

The group resolved to establish Bourbaki's existence irrefutably by publishing a note under his name in the *Comptes-Rendus* (Proceedings) of the French Academy of Sciences. They needed a first name for Bourbaki, having originally decided that their society be named after the general from Husson's prank at the École Normale, since there had already been a published result under that name in Kosambi's paper inspired by Weil.

Eveline de Possel, the wife of one of the members, was present at the meetings and it was decided that she should be Bourbaki's godmother. In a ceremony they held, Eveline baptized the baby Nicholas.[16] The note written in his name was indeed published in the *Comptes-Rendus* of the French Academy of Sciences, and Nicolas Bourbaki thus became a mathematician with a publication to his name. The group also invented a daughter for Nicolas Bourbaki, whom they named Betti and even printed invitation cards for her wedding. Nicolas Bourbaki was to be a person with relatives and academic publications and a personal history.

It was around this time that Weil, the head of the group,

and Eveline de Possel became involved with each other. Eventually, she would divorce her husband and marry Weil. But even before her divorce was finalized, at the beginning of 1936, Weil took her on a trip to Spain and got her drunk so she could watch a bullfight in Seville without tears.[17]

On their way back to France, the couple passed through the Escorial, and Weil was so dazzled by the look of these impressive buildings that he got the idea to arrange to have the summer meeting of Bourbaki at this location—in a high school near the monastery, which often provided lodgings for academic guests. But in July of that year, the Spanish Civil War broke out and these plans had to be canceled. As the war began, André's sister, Simone Weil, left for Barcelona and from there continued on to the front in Aragon, to help the Republican forces. At about the same time, Alexandre Grothendieck's parents were also off to Spain to help the same side in the war.

In September 1936, Bourbaki convened for the "Escorial congress" in an alternative location, in Chançay, in the French region of Touraine. Here, Claude Chevalley's family owned a beautiful house, which Claude's mother gladly offered as the location for the meetings. Two Bourbaki congresses would take place at the Chevalley family property.

Reminiscing about these meetings at his parents' estate, Claude Chevalley said, "The Bourbakis arrived at the station at Amboise, and those who were already there let out a frightful howl: 'Bourbaki! Bourbaki!' You would have taken us for a band of madmen. There, that was the Bourbaki style!"[18]

Chevalley continued: "Strong bonds of friendship existed between us; and when the problem of recruiting new members was raised, we were all in agreement that members should be chosen as much for their social manner as for their mathematical ability. This allowed our work to submit to a rule of unanimity: anyone had the right to impose a veto. As a general rule, unanimity over a text only appeared at the end of seven or eight successive drafts. When a draft was rejected, there was a procedure foreseen for its improvement. 'The Tribe,' a report of the congress, related the discussions and decisions on the subject."[19]

At this congress and the following ones, the Bourbaki group changed its working methods. For each topic, a general discussion took place. Bourbaki always argued among themselves, and the meetings were never quiet or orderly. This, in fact, became the group's hallmark. And one of their innovations was to do mathematics in the open air—on lawns and in nature. Traditionally, mathematics was a dry discipline, taught with chalk on a blackboard in a stuffy room. Bourbaki immediately broke this mold, pursuing mathematics in attractive locations in nature, and never in a stifling room. The members wanted to enjoy life, and to enliven mathematics.

Once a discussion was over, a writer was designated. This writer produced a preliminary draft of the topic, and then the group discussed the draft and argued about its various points. Another writer was then designated to write the second draft, after which a new discussion would take place, and so on, until the final draft was produced. Given this work system, it was impossible ever to be able to attribute any given text produced by Bourbaki to any particular person in the group. All decisions had to be unanimous, and any decision could be challenged at any time. This made for a very cumbersome process of writing, but it did ensure that the final product was always created by a group rather than by any particular individual.

Describing this process, Weil wrote in his memoirs: "No doubt it required a major act of faith to think that this process would produce results, but we had faith in Bourbaki."[20] In reality, it so happened that Jean Dieudonné ended up writing most of the final drafts of the product of Bourbaki's work.[21] He was an excellent writer, and he had a very good style.

The final consensus necessary to produce a text by Bourbaki surprised even the members of the group. It would happen as if by magic after many hours of arguing and discussing every small detail. Perhaps there is a lesson here for academics in other areas as well as people in business or government: when there is a will, there is always a way of reaching a final agreement.

At the "Escorial congress," the first text produced by Bourbaki was read and adopted in its final form. This was the text of set theory, the original "Abstract Package." Set theory is viewed as the foundation of all of mathematics, and it was Bourbaki's great idea to devote the first book produced by the group to this key foundational topic. André Weil, in fact, invented for this book the universal notation we use today for the empty set: $\varnothing$.

This symbol comes from the Norwegian alphabet, which Weil had encountered in his travels. The Bourbaki text on set theory thus introduced this new mathematical symbol. It also introduced a whole new way of thinking about mathematics: placing sets and operations on sets at the base of all of mathematics. This would lead to the New Math that schools throughout the world began to teach in the 1950s.

The mathematicians of Bourbaki realized the importance of what they were trying to achieve. "I absolutely had the impression of bringing light into the world—the mathematical world, you understand," Claude Chevalley told the French publication *Dédales* when he was interviewed about the early years of Bourbaki in 1985.[22]

Apparently, the members of Bourbaki were becoming all too aware of their importance, and of their influence on the world of mathematics. Chevalley continued: "It went hand in hand with the absolute certainty of our superiority over other mathematicians—a certainty that we held something of a higher level than the rest of mathematics of the day."[23]

Perhaps it was this arrogance in their point of view that would lead to the eventual decline of the group, for there were certainly excellent mathematicians that were outside Bourbaki, and even outside France. Mathematics continued its progress and would be developed further, beyond Bourbaki's dream, in Germany, Japan, the UK, and the United States, among others.

Perhaps Chevalley himself was aware of the futility of the group's euphoria. He described Bourbaki's process of the making of a mathematical text: "It is more than just improving it. It is to treat it according to the norms which Bourbaki wanted to introduce into mathematics, essentially the theory of sets and the notion of structure. It is the notion of structure that is truly Bourbakian. But with this feeling of accomplishing a gigantic task we came to the certainty that it would be impossible to achieve."[24]

As history would prove, the structuralist approach promoted by Bourbaki was innovative and extremely fruitful, but basing the entire edifice of modern mathematics on the theory of sets would not prove to be the best decision. Eventually, this choice—logical as it may have seemed at the time—would become the group's Achilles' heel.

The "Escorial congress" led to other important work and established norms for the group and its work. Each book, it was decided, should have not only problems for solution by the reader, but also a historical appendix. This decision, too, was influenced by a suggestion made by André Weil, who had always been interested in the history of mathematics and studied it throughout his life. Eventually, one of Bourbaki's most important books would be its volume on the history of mathematics.

Bourbaki set as one of its main tasks during this meeting to concentrate on the idea of *structure* in mathematics. Weil wrote in his memoirs that he was aware at the time that the idea of structure had already entered the field of linguistics.[25] Through his association with linguist Emile Benveniste, Weil was aware of the structuralist work done so successfully by Russian linguist Roman Jakobson. The Bourbaki mathematicians wanted to adapt this new viewpoint in science to their own discipline.

In late 1937, Eveline's divorce was finalized and she and André Weil were making plans to marry in October of that year. Meanwhile, Bourbaki met again in Chançay in September for a new congress. Simone Weil was there, too, hoping to improve her knowledge of mathematics. The mathematicians met in the country surroundings, and this time worked on their next books, Bourbaki's texts on general topology and topological vector spaces. Topology was a natural place to continue from set theory since the theory's basic elements are, indeed, sets—open sets and closed sets, upon which an entire area of mathematics is built. Topological vector spaces have the additional structure of vectors added to the topology.

During this congress, Bourbaki took its first political action, albeit on a small scale. The French physicist and Nobel Prize winner, Jean Perrin, who was undersecretary of state for scientific research, had proposed a system of medals to be awarded to scientists, with increasing monetary value. The announcement of his plan was made while Bourbaki was meeting in Chançay, and there was unanimous opposition to the plan by all present, as they felt that the prizes would corrupt scientists. The group organized a petition against the plan, and obtained 400 signatures, which they presented to the Minister of Education, resulting in the defeat of the initiative.[26]

Many of the members of Bourbki, including André Weil, were reserve officers, since, traditionally, graduates of the École Normale were expected to serve as officers. As the clouds of war were gathering over Europe in the years after Hitler took power in Germany, it became clear to these reserve officers that they would be called to serve their country once it became probable that war would break out. André Weil, however, was determined not to serve. In his memoirs, he is clear about the fact that he was not a conscientious objector. Killing a fly "does not fill me with remorse," he wrote.[27] He simply did not want to serve in the army. It was not his dharma, he wrote, and continued on a long discourse quoting Kant and the *Gita* to explain to the reader his decision not to serve his country in time of need.

In 1938, Bourbaki convened for its new congress in Dieulefit. The congress took place exactly at the time of the infamous Munich conference. "There was a sinister foreboding in the air," Weil wrote.[28] The participants did very little mathematics. Instead, they spent hours huddled around the radio listening to reports of the Munich meeting and worried about the imminent threat of war. They devoured the newspapers every morning, looking for hints of what might transpire. In the middle of the congress, fearing that war might break out any moment and he would be called up for reserve duty,

Weil abandoned his colleagues on "some pretext or another and left for Switzerland."[29] Before doing so, he consulted with Delsarte, who was also a reserve officer but who had no plans of leaving France. After two days in Switzerland, it became clear to Weil that war would be averted for the moment, and he returned to the congress. Then the congress was over, and little had been achieved.

It became clear again that Hitler would not be dissuaded from his designs on the continent of Europe, and that Chamberlain's "Peace in our time" was merely an illusion. Weil then escaped to Holland. From there he went to Finland, was arrested, and eventually would have to serve his country, anyway—as a private rather than an officer.

While André Weil was still in wartime France in early 1940, waiting for a Vichy passport that would allow him to escape to the United States, the members of Bourbaki were all very concerned about sustaining their group despite the hardships of war. The 1938 congress at Dieulefit had been the last one held. Bourbaki's bulletin, "The Tribe," had been put together in March 1940 and distributed to the membership, and a second part of the textbook they had put together in earlier congresses was about to be published in late 1940.

The group now planned to hold a meeting despite the difficulties, and called a new congress for the fall of 1940. This Bourbaki congress took place in Clermont-Ferrand, the only place to which all the members could come, given that France was now sliced into two parts. At this congress, Laurent Schwartz was inducted as a new member of the group. He was the only member to join Bourbaki during the war. After Weil's departure for the U.S., the members of Bourbaki who remained in Europe managed to keep the group alive and indeed to meet from time to time, even without the group's main founder's participation. After the war, he would come over from America to participate in new congresses.

Jean Delsarte, one of the founders of Bourbaki, was a captain in the French army during the war, and found himself leading French troops in a humiliating retreat from the attacking Germans. Their retreat led them along a familiar route, as the history of French defeat by the Germans was repeating itself for the third time. As the army, in disarray, was escaping toward the Swiss border just as it had done when fleeing the Prussians in 1870, Delsarte overheard one of his soldiers comment: "We are the army of Bourbaki . . ."

eight

THE ACHIEVEMENTS OF BOURBAKI

ACCORDING TO THE French historian of mathematics Denis Guedj:

> Animated by a profound faith in the unity of mathematics, and wishing to be 'universal mathematicians', Bourbaki undertook to derive the whole of the mathematical universe from a single starting point.

That starting point was the theory of sets.

Twenty-five centuries earlier, Euclid based his entire *Elements* on the elementary notions of point, line, circle, and other geometrical concepts, and on this foundation he constructed his entire system of mathematics. Bourbaki wanted to found their own system of mathematics in a similar way to that of Euclid. Their stated aim was to produce a new set of Euclid's *elements*—"to last for the next 2,000 years." In that spirit, they named their own series of books they planned to publish, "Elements de mathématique." They purposely spelled mathematics in the singular, in French: mathematique, rather than the usual way, mathematiques. They chose this innovative new word to stress the *unity* of mathematics. They were striving to write their series of books, their *elements,* to be "prepared according to the axiomatic methods, always maintaining as a horizon the possibility of a total formalization, our Treatise aims at a perfect rigor."[3]

Guedj notes that the Bourbaki group was not the first in history to have entertained such an ambitious project, but that they were the only ones to have advanced so far toward the realization of this goal. To carry on toward their goal, the Bourbaki group chose two powerful methods. One was the idea of *axiomatization*; and the other was the general notion of *structure*. Axiomatization was an idea they drew directly from Euclid, as further enhanced by the German mathematician David Hilbert and others. But the second idea, the important concept of structure as applied to mathematics, was something that Bourbaki had to invent. "It counts as one of the most beautiful jewels of twentieth-century mathematics."[4]

According to Claude Chevalley, another innovation by Bourbaki was "the principle that every fact in mathematics must have an explanation." This principle is separate from the idea of causality, meaning that one fact causes the occurrence of another. Bourbaki held that "anything that was purely the result of a calculation was not considered by us to be a good proof."[5]

All of the principles mentioned above are extremely useful and important in mathematics, and we could hardly imagine modern mathematics today without an explanation for every fact, an axiomatic foundation for the discipline, and the underlying concept of structure. This last principle, however, is unique in its importance and has surpassed all others. The reason for this is that the principle of structure, which, as we know, originated in linguistics, has importance far beyond the confines of mathematics. The concept is, at its core, a mathematical one; and Bourbaki has brought it into the forefront of mathematical thinking. But the concept is so powerful and so fundamental to science and to the processes of human thinking that it found applications in virtually every area of human interest. The concept of structure, in fact, started an entire revolution in human thinking and in human philosophy. Bourbaki was the key proponent of this concept and made it into the universal core of human thought.

What are the mathematical essentials of the concept of structure? Bourbaki developed a number of "mother structures," as the group called them. Two of them were the notion of nearness and the structure of a group. The idea of nearness, or neighborhood, comes from the mathematical field of topology. Here, we study the notion of neighborhood. There are precise mathematical notions that define "nearness" of two points, and Bourbaki made these notions of paramount importance in defining nearness. These ideas are abstract, and yet they have found applications in psychology and the theory of learning, for example. A psychologist may want to know whether two concepts, such as a curve and a straight line, are "near" each other in the human mind. Here, quasi-mathematical notions based on the topological idea of nearness may be used.

The idea of the structure of a group was one that held special importance for Bourbaki, since the area in which it resides, abstract algebra, was the favorite research area of several of its members. The idea of a group was over a hundred years old when Bourbaki tackled it. It had originated in the late eighteenth century and was developed well by the young Evariste Galois around 1830, just before this twenty-year-old genius died tragically in a duel. If we look at the set of numbers 123, their various possible orders form a *group*. It is called the *permutation group of n elements* (in this example, n=3). The members of this group are: 123, 132, 213, 231, 312, and 321. The group thus has six distinct elements. But this group appears in many other guises.

The idea of an abstract group is that *it does not matter what the actual situation is, or the application from which the group arises. What matters is only the inherent structure of the group.*

Bourbaki attached great importance to this idea of structure. The internal nature of the group above—completely independently of where this group came from, whether it is triangles, or numbers, or solutions of equations—was the key element, the *structure,* that Bourbaki cared about. Structure could thus be seen as a latent *code* or symbolism for what was "going on" mathematically in a given situation. And the situation itself no longer mattered to Bourbaki—it was only the code, the symbolism, the latent structure that was of interest to these mathematicians.

Bourbaki did not invent the idea of structure: it existed, in the nonmathematical setting of linguistics. And soon, while Bourbaki was working on its structures, the idea would transform itself into anthropology, and from there to psychology, and eventually—again through linguistics—into literature, an area in which one could hardly imagine a mathematical idea could find fertile ground. But, in fact, within a few years, the idea of structure would dominate all thinking in Western culture. And Bourbaki would be a major force in this innovation, enabling social scientists, humanists, and writers to use and understand this immensely important concept.

Sometimes an idea that has been developed earlier waits until the right person grasps it and makes it a reality. This is what would happen with the idea of structure in 1942. The great anthropologist Claude Lévi-Strauss would attempt to apply structure, learned from the linguist Roman Jakobson, in anthropology. But he would need the mathematical underpinnings of structure—and these would be supplied to him by Bourbaki. A meeting in New York that year between Lévi-Strauss and André Weil would result in the solution of a difficult problem of kinship studied by Lévi-Strauss.

André Weil would use structure—and, in fact, the structure of a mathematical group as described above—to solve Lévi-Strauss's problem. In doing so, Weil would enable the transfer of the mathematical idea of structure into anthropology, and from it into other fields. Bourbaki would thus be instrumental in unleashing the powerful concept of structure. Our culture would never be the same after this important development.

Sometimes, several latent structures can underlie a mathematical situation. This is the case with topological groups, the subject of great interest to Grothendieck in his early career, as well as to other French mathematicians. One structure in topological groups is the "nearness" structure of topology, and the other is the structure of a group. Another example of dual structures is the area of topological vector spaces. A vector, in its simplest description, is an arrow with a length and a direction.

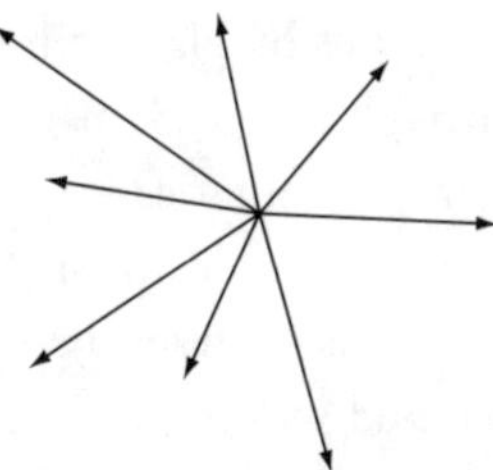

A vector space is thus a space that is defined by vectors. The topology of a space, on the other hand, is a system of open sets (which define the notion of "nearness"). The picture below shows the dual structure of a topological vector space: a space defined both by a collection of open sets—in this case, balls of various sizes—and by the vectors.

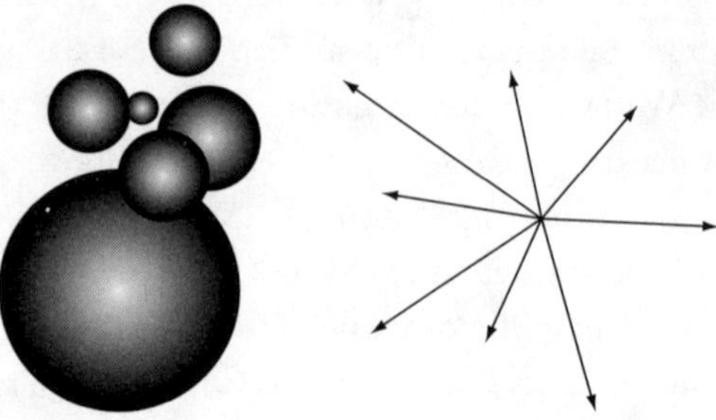

The concept of structure thus allowed the mathematicians of Bourbaki to develop complicated mathematics based on elementary concepts and to construct the whole edifice of modern mathematics on certain basic notions. These same notions would find application in a variety of fields.

FOLLOWING THE END of the war, the secret mathematical society of Nicolas Bourbaki flourished and assumed a key role in the development of mathematical ideas. Bourbaki began to transform mathematics by placing it on a firm theoretical foundation. Through the work of Bourbaki, the French gained prominence in mathematics as German mathematics, dominant in the prewar years, declined. There took place the birth of the New Math, based on set theory, which dominated American education and was important in the educational systems of other nations for a period of time. This approach owes its inception to the work of Bourbaki. What were the achievements of this unique group of mathematicians?

An article by Marjorie Senechal in the *Mathematical*

Intelligencer in 1998, featuring an interview with the renowned former key member of Bourbaki, Pierre Cartier, summarized the achievements of the group as follows:[6]

Bourbaki was the discoverer (or inventor) of the notion of mathematical structure.

Bourbaki was one of the greatest abstractionist movements of the twentieth century.

While Bourbaki was a small group of mathematicians, it exerted enormous influence on the development of mathematics in the twentieth century.

Pierre Cartier was born in Sedan, in northern France near the Belgian border, in 1932. He served as a midshipman on a French naval ship during the Algerian war, mostly patrolling the waters off the Algerian coast. Later, because of his mathematical skills and education, he was assigned to teach mathematics to naval personnel. Cartier's interests lay both in pure mathematics and in mathematical physics.[7]

Cartier was accepted to the École Normale Supérieure in Paris, and was a student of Henri Cartan. Because he was a very bright student and showed great promise even as a first-year student, Cartan requested that Bourbaki invite Cartier to their 1951 meeting in Pervoux in the French Alps. During this meeting, Bourbaki was working on editing its book on Lie groups.

The important modern topic of Lie groups—a key tool in theoretical physics in addition to its roles in mathematics—was the latest topic of the books written by Bourbaki. By that time, the group had produced its groundbreaking books on the following topics (with titles translated into English):

Set Theory
Algebra

Topology
Functions of One Real Variable
Topological Vector Spaces
Integration

There was also a short book, a summary of results only, on differential and analytical varieties that dealt with the theory of manifolds. The set theory book, which was also intended as a summary of results, had grown into the first full text written by the group. It set the stage for the development of all of mathematics based on the notions of a set, set membership, inclusion, and the elementary set operations of union, intersection, and symmetric difference. Set theory has deep paradoxes and inconsistencies inherent in its structure—which became apparent in the early part of the twentieth century through the work of Georg Cantor, Kurt Gödel, and Bertrand Russell. But the Bourbaki group had apparently decided to ignore these theoretical difficulties, and to base all of mathematics as practiced by its members, and inherent in its writings, on the foundation of the theory of sets.

The topic studied at Pelvoux, Lie groups, was at the other end of the spectrum from set theory. While the theory of sets forms the rudimentary foundation of the mathematics practiced by Bourbaki and others, Lie groups are among the most complex and involved topics in mathematics—occupying a place at the intersection of the continuous world of topology and analysis and the discrete world of algebra.

Laurent Schwartz, who had joined Bourbaki during the Second World War while hiding from the Nazis, was one of the members who were most active at the Pelvoux meeting. By then Schwartz had made his important discovery of distributions, which made him famous. According to Pierre

Cartier, the mathematics students at the École Normale were all students of Cartan and Schwartz, and they all strove to understand distributions and their role in the theory of Lie groups. For Cartier, being invited as a nineteen-year-old student to attend a meeting of Bourbaki was an incredible opportunity to learn about this important topic "from the inside."[8] Eventually, he would do important work himself in Lie group theory.

Cartier benefited greatly from the opportunity to learn from the prominent mathematicians present at the congress, some of whom he had seen only from a distance. At this meeting he learned about the unique way in which Bourbaki did mathematics: the work on the textbooks, research work, the discussions of theorems and ideas, and the work of students, were all interlinked. According to Cartier, it was this integration by Bourbaki of ideas, research, and writing that accounted, in part, for the great success of French mathematics in the 1950s and 1960s.

For Cartier, the meeting was his first real exposure to modern mathematics. He had come from a small city, one that had suffered much during the Second World War and had not yet recovered. He described his school as provincial and outdated in its programs; some of his teachers were good, but they were "far away from modern science."[9] Cartier explains how Bourbaki materialized in a virtual intellectual vacuum in postwar France. For example, he said, the geometry he was taught at his school was "classical geometry, in the uncultivated, synthetic way." The mathematics and physics he was taught were "totally outmoded at that time—totally." He remembers, for example, that in a course called General Physics at the Sorbonne, the professor made the statement, "Gentlemen [there were ladies in the audience as well], in my class, what some people call the 'atomic hypothesis' has no place."

Cartier notes that this happened in 1950, five years after the atomic bomb was dropped on Hiroshima. "This shows," he adds, "what the French university was like at that time. In order to understand the influence of Bourbaki, you have to understand that. Bourbaki came into a vacuum. . . . Obviously, in the early fifties, the teaching of science was very poor. It took Bourbaki about five or six years to subvert the whole system. By 1957 or 1958, the subversion had almost been complete in Paris."[10] In the 1950s, Bourbaki published one or two volumes every year, and mathematics students rushed to the bookstores to purchase these books and finally learn mathematics in a rigorous and comprehensive and accurate way. During this extremely productive period, Bourbaki met three times a year: a weeklong meeting took place in the fall and another in the spring, and there was a two-week-long congress each summer. At these meetings, the members of Bourbaki worked very hard—often for as long as ten or twelve hours a day. And they produced voluminous works: the published books of Bourbaki comprise 10,000 pages. This meant that 1,000 to 2,000 pages of preliminary reports and drafts were written by the group every year.[11]

Pierre Cartier was officially accepted as a member of Bourbaki in 1955, and he retired from the group almost thirty years later, in 1983, when he passed the age of fifty—the mandatory retirement age for the members. He took part in the most important and fecund years of the group's work. According to his own estimate, he contributed a total of 200 pages a year to Bourbaki's effort.

Bourbaki was always a very small group of mathematicians, typically numbering about twelve people. Its first generation was that of the founding fathers, those who created the group in 1934: Weil, Cartan, Chevalley, Delsarte, de Possel, and Dieudonné. Others joined the group, and others left its ranks, so that some years later there were about twelve members, and that number remained roughly constant. Laurent Schwartz was the only mathematician to join Bourbaki during the war, so his is considered an intermediate generation. After the war, a number of members joined: Jean-Pierre Serre, Pierre Samuel, Jean-Louis Kozul, Jacques Dixmier, Roger Godement, and Sammy Eilenberg. These people constituted the second generation of Bourbaki. In the 1950s, the third generation of mathematicians joined Bourbaki. These people included Alexandre Grothendieck, François Bruhat, Serge Lang, the American mathematician John Tate, Pierre Cartier, and the Swiss mathematician Armand Borel.[12]

The Bourbaki Seminar was begun by the group after the war, and had a much wider participation than that of the members of Bourbaki. In fact, many of the participants in this regular seminar had no connection at all with the group named Bourbaki. The seminar was an active outlet for research results in mathematics in France: here papers on important mathematical topics were presented and discussed. This seminar thrived in the 1950s and 1960s, in part because there was no French mathematical society that would organize such talks. Typically, members of Bourbaki ran the seminars, and the seminar became the mouthpiece of the organization. Thus Bourbaki communicated with the rest of the mathematical world through its books and through ideas disseminated in its seminars. But as the venue grew, mathematicians would present papers in these seminars that had little or nothing to do with Bourbaki. The Bourbaki Seminar became a good venue for French and other mathematicians to present and discuss their work. In those days, about 200 people attended each seminar. The seminar continues to this day, and takes place in a lecture hall at the Institut Henri Poincaré in Paris. Because there is now an active French mathematical society, the Bourbaki Seminar is now somewhat less heavily attended. Typically, less than seventy mathematicians attend each meeting.[13]

nine

BOURBAKI'S REIGN

BOURBAKI'S OEUVRE WAS conceived as comprising two distinct parts. The first part was purely foundational, in which the edifice of modern mathematics was constructed starting with set theory, then continuing to the beautiful structures of abstract algebra, following on to topology, continuing to the area of the theory behind the calculus (called mathematical analysis), followed by topological vector spaces, and ending with Lebesgue integration.

The last four of these books gives the foundations of analysis as seen through the eyes of Bourbaki, with an emphasis on functional analysis. The second part of the work of Bourbaki is an exposition of complicated, involved topics in modern mathematics: treatises on Lie groups and commutative algebra. This part enjoyed unusual breadth and depth in its treatment by Bourbaki, thanks mostly to the fact that the world's leading experts in these areas at that time were members of the group.[1]

Generally, Bourbaki avoided almost every kind of visual illustration in its works: there are no tables or figures accompanying the text—with the exception of the symbol "dangerous curves," which has become Bourbaki's hallmark, indicating in the margin of the text that a passage may be difficult. But the volumes in the second part of Bourbaki's work, the ones on Lie groups and commutative algebra, are unusual in this respect: they do contain tables and visual illustrations. This aspect is due, according to Pierre Cartier, to the influence of one man: Armand Borel. Borel claimed that this need for pictures was part of the Swiss national character, and would often say in Bourbaki's discussions, "I'm the Swiss peasant."[2]

Armand Borel (1923–2003) was born in La Chaux-de-Fonds, Switzerland. When he was a student at the Swiss Federal Institute of Technology (ETH) in Zurich he read carefully the Bourbaki texts and learned much mathematics from them, but he disliked many of the features of these books. They seemed to strive for excessive generality, there was no apparent consideration for the reader, and there was an internal reference system with a total absence of any references to works by mathematicians outside of the Bourbaki group. This struck him as very strange. There was in these books also an emphasis on generality for its own sake—with no specific problems to illustrate the concepts.[3]

In 1949, Borel won a fellowship from the French Centre Nationale de la Recherche Scientifique (CNRS), which allowed him to go to Paris. There, the young mathematician became acquainted with key members of the group: Henri Cartan, Jean Dieudonné, and Laurent Schwartz. He also met some of the younger members of Bourbaki, including Roger Godement, Pierre Samuel, Jacques Dixmier, and Jean-Pierre Serre. This led to important discussions of mathematics Borel had with many members of the group, and to strong friendships.

Armand Borel also attended the Bourbaki Seminars, at which new developments in mathematics were being presented and discussed.[4] By that time, Jean-Pierre Serre had emerged as the new leader of the group. New leadership was necessary because André Weil had remained in the United States even after the war, taking various academic positions, including one at the University of Chicago, as well as his final position as head of the mathematics group at the Institute for Advanced Study at Princeton. Weil was thus mostly away from the rest of the Bourbaki "tribe," as it liked to call itself, and thus less effective in his interactions with the rest of the other members.

But Serre, a second-generation member of Bourbaki, is one of the most gifted mathematicians alive today, and was the youngest recipient of the Fields Medal—the "Nobel Prize of mathematics." But as a leader, he probably left something to be desired. Serre developed a strong friendship with Alexandre Grothendieck, who also joined the Bourbaki group after the war, and a lengthy correspondence developed between the two men, a correspondence about mathematical topics that has now been published.[5] However, to act as a charismatic leader of a group that was rewriting the history of mathematics was another role, one to which Armand Borel would have been better suited than Serre.

Borel, who died a few years ago, was an excellent mathematician who also liked to work with people and to socialize and enjoy the company of others. He organized many events, among them "Armand Borel's Camp," in which he taught number theory to young mathematicians in a summer camp in Oregon in 1976. Borel contributed much to the Bourbaki projects, and he is sorely missed by those who knew him. As soon as he was introduced to Bourbaki as a young mathematician, Borel's interactions with the members of the group completely changed the opinion he had formed about them from reading their books. Instead of the closed-minded group of individuals who cared only about their goals of abstraction and generality that he had imagined them to be, Borel now found in his new friends people who had a rather broad outlook on life and on mathematics in particular. "They knew so much, and knew it so well," he wrote, that they could "go to the essential points and reformulate mathematics in a more comprehensive and conceptual way. Even when discussing a topic more familiar to me than to them, their sharp questions often gave me the impression I had not really thought it through."[6]

Borel was smitten with this amazing group of mathematicians and was eager to participate in the creation of new mathematical ideas. Sometime later, he was invited to take part in a Bourbaki congress. Here, he was bewildered: this meeting seemed like a private affair in which only the books the group was writing at the time were discussed and analyzed and dissected and argued about. Discussions often turned into chaotic shouting matches in which an obscure point in a text was fought over by various individuals. People would shout at high decibel levels, sometimes three or four people at the same time. Some who were invited to attend a meeting but were not members of the group would come out with the impression that a Bourbaki congress was "a gathering of madmen."[7] Only later, when Borel was invited to become a member of the group, did he begin to understand that the apparent chaos of the system was there by design. Quoting André Weil's description of Bourbaki's style of discussion and work at congresses, Borel wrote that Bourbaki was "keeping in our discussions a carefully disorganized character. In a meeting of the group, there has never been a president. Anyone speaks who wants to, and everyone has the right to interrupt him. The anarchic character of these discussions has been maintained throughout the existence of the group."[8] It was, in fact, the apparent chaos that made Bourbaki so effective. For through these very open discussions any flaw in the manuscripts would be detected and corrected, and the final product was truly the work of a group rather than of an individual writer.

There were disagreements and power struggles within the group. Claude Chevalley reminisced: "There was a terrific struggle for years between Weil and de Possel."[9] The disagreements eventually led to a great resentment by some members of the group toward others. De Possel left the group at some point, and so did Serge Lang, who moved to the United States. Ehresmann attended all meetings, and then, from one day to the next, he stopped coming.[10] Chevally, too, began to draw away from the group at some point. "My drawing apart from Bourbaki," he recounted, "was progressive. The final straw was the position taken by Dieudonné in 1968." This position had to do with Bourbaki getting involved with university politics in France, "a cleaning-out of the university," as it was called.[11]

In spite of all the difficulties inherent in the modus operandi of Bourbaki, the volumes of its work kept appearing. This seemed to surprise even the founding members. Apparently the chaotic system of producing mathematical writings was working all too well in a rather mysterious way. According to Armand Borel, one reason for this unexpectedly high level of success of Bourbaki was the unflinching commitment of its members. All the members of Bourbaki shared a strong belief in the worthiness of the project they were undertaking, however distant the goals may have seemed. They were all willing to devote an inordinate amount of time and effort to the project of producing the Bourbaki works—despite the grueling hours. Long drafts were written and discussed and rewritten, and this took a tremendous amount of energy and commitment. Often a draft worked over for many days would be heavily criticized once it came up for general discussion by the group; sometimes, the draft was even dismissed out of hand, or rejected after a reading of only a few pages. Many manuscripts, even when read with interest, ended up not being published. The manuscripts that did get published after this grueling process brought no personal recognition to the people who wrote them. Their author was Nicolas Bourbaki. This was "altogether a truly unselfish, anonymous, demanding work by people striving to give the best possible exposition of basic mathematics, moved by their belief in its unity and ultimate simplicity."[12]

According to Borel, another reason for the success of the enterprise was the "superhuman efficiency of Dieudonné." Jean Dieudonné was the major writer of the final versions of the manuscripts. For twenty-five years, Dieudonné would start each morning by writing a few pages for Bourbaki. His role in writing the final versions of the books accounts in large part for the uniformity of the writing style of the Bourbaki volumes. This style, however, was not Dieudonné's personal style, but rather a style he had adopted purposely for the writing of the Bourbaki books.

The hard work done so selflessly by the highly committed members of Bourbaki on producing their oeuvre—"a new version of Euclid to last for the next 2,000 years"—did bear fruit. The 1950s were a period of spreading influence of Bourbaki worldwide. In particular, the group has been responsible for the "French explosion" in the field of algebraic topology, an area of mathematics in which a number of leading French mathematicians made important advances.

Algebraic topology is the kind of topology familiar to many nonmathematicians. This field has to do with algebraic properties of curves and surfaces, such as the familiar notion of the genus of a surface—the number of holes it contains. Many laypersons are familiar with this notion, knowing that the genus of a doughnut is one, because a doughnut has one hole; the genus of a coffee cup with two handles is two, because it has two holes, and so on. Of course, algebraic topology is far more complicated than this example; the subject is an area at the crossroads of topology and algebra, and thus it can lead to powerful mathematical results. Here, the *structure* of a problem in topology can become apparent by the use of algebraic methods—for example, by identifying a *group* that underlies the topological structure. Thus abstract algebra is used as a tool for uncovering truths about topological surfaces, truths that would not otherwise be apparent, or amenable to analysis.

During these fruitful years of the group, the 1950s and 1960s, Bourbaki also enjoyed great successes in expanding our understanding of coherent sheaves in analytic geometry, and it made great strides in algebraic geometry.[13] Algebraic geometry is an area in which the geometry of numbers is studied. This area has important applications in number theory, as well as in other fields. Here, too, French mathematicians made great advances in this field. Prime among them was Alexandre Grothendieck, who almost single-handedly rewrote much of algebraic geometry. Borel wrote about these successes of Bourbaki: "We had witnessed progress in, and a unification of, a big chunk of mathematics."[14]

The first generation of Bourbaki had learned mathematics the old-fashioned way, meaning that the teachers of that generation were still not rigorous in their approach: proofs lacked detail or were simply wrong, and a foundation for mathematics and the idea of structure were completely missing. The charge of that generation was to redo mathematics and to completely rewrite the way mathematics should be pursued and taught.

The second generation of Bourbaki consisted of people who had already been exposed to the new ideas. These people knew that rigor, generality, abstraction, and structure were of paramount importance in mathematics. Their charge was to continue the good work: to publish more books using the new approach, and to conduct research and teach mathematics using these important ideas forged by the first generation.

The third generation of Bourbaki, the one that included Pierre Cartier, Armand Borel, and Alexandre Grothendieck, did not have to prove that the new way of doing mathematics was better than the old one, because they had already been raised as mathematicians in the new way of thinking. Bourbaki's structuralism, and its emphasis on rigor and abstraction, had already triumphed. Some of these people had been taught in the old and ineffectual way of doing mathematics when they were in high school, but at the École Normale and other universities in Paris these mathematicians were taught mathematics the Bourbaki way.

This third generation of Bourbaki, therefore, was not dogmatic and did not feel it had to prove anything. The mathematicians in this generation only had to continue the work of the previous two generations of members of Bourbaki. Bourbaki's achievements were already very evident, and the Bourbaki method had won over in France. Worldwide, mathematicians were paying very close attention to the advances made by Bourbaki. They were studying the new methods and producing new mathematical results using the novel approaches pioneered by the French group. The New

Math, a way of teaching mathematics using the ideas of sets, was becoming the main method of teaching mathematical ideas in high schools in America and elsewhere.

THE MEMBERS OF the Bourbaki group recognized only one French mathematician of the pre-Bourbaki era as their godfather: the geometer Elie Cartan, Henri's father. Other French mathematicians of the generation before Bourbaki were not well liked, including Lebesgue and Poincaré.

During the 1950s, when Pierre Cartier joined Bourbaki, it was not fashionable to value the work of Henri Poincaré, who was one of the most important French mathematicians of all time. Poincaré was called "the last universalist." He was a mathematician who understood a great deal about a very wide range of mathematical topics. Poincaré derived very important results in mathematics, and a conjecture he had made a century ago now seems to have been proved by contemporary Russian mathematician Grigori Perelman. But Bourbaki viewed Poincaré's way of doing mathematics as old-fashioned. Poincaré's style of doing mathematics and the style of Bourbaki were at odds with each other.[15] Poincaré was intuitive and detail-oriented. To him, mathematics was an art. He did not care about structures or axiomatics. He cared about seeing things clearly and producing results—in whatever way he could.

In the 1960s, a new generation joined the group: the fourth generation of Bourbaki. These were mostly former students of Alexandre Grothendieck. Grothendieck himself was a member of Bourbaki for about ten years, and contributed greatly to the development of modern mathematics. But he came to a conflict with the group, and left it in anger. Before he left, there were frequent clashes within the group, as well as generational conflicts.[16]

As Armand Borel wrote, "The fifties also saw the emergence of someone who was even more of an incarnation of Bourbaki in his quest for the most powerful, most general, and most basic—namely, Alexandre Grothendieck."[17] Grothendieck "quickly made mincemeat of many problems on topological vector spaces put to him by Dieudonné and Schwartz, and proceeded to establish a far-reaching theory. Then he turned his attention to algebraic topology, analytic and algebraic geometry, and soon came up with a version of the Riemann-Roch theorem that took everyone by surprise, already by its formulation, steeped in functorial thinking, way ahead of anyone else. As major as it was, it turned out to be just the beginning of his fundamental work in algebraic geometry."[18]

Grothendieck was unhappy with Bourbaki because he felt their work and aims were not ambitious enough. Eventually, he would write his own series of books, and leave Bourbaki. By the fourth generation, the goal of the group was not as clear as it had been earlier. Grothendieck had by then developed his own, more ambitious, program outside of Bourbaki. Thus the need for Bourbaki was less obvious, and a lack of a global understanding of mathematics was beginning to manifest itself vis-à-vis these new developments.[19] The members of the group had become more specialized in their interests, and less able to see the larger picture.

BUT AS GREAT success came, almost overnight as it may have seemed, to the members of Bourbaki—and with it, invitations for talks and presentations around the world, prizes, and recognition for their collective achievements—there also arose a degree of resentment. Not every mathematician was happy with the new approach to mathematics. Many around the world felt that Bourbaki had gone too far, that the group was now pursuing generality for generality's sake, and that they cared less about explaining mathematics than they cared about presenting results that were very abstract. This negative trend would haunt the group in the years to come, as the criticism spread to the various methods employed by the group.

One major criticism has been that Bourbaki was overly formal, too abstract, and much more rigorous than necessary, thus making it unnecessarily difficult to read and understand mathematics, and to use it in a meaningful way. Bourbaki's emphasis on precise definitions has found disapproval in the international mathematical community.[20] But Bourbaki's main aim had been to improve and deepen the understanding of mathematical concepts, not to make them obscure, thus the criticism was somewhat justified. For the question arose: How far should generality and abstraction go before they become an impediment to understanding? The proofs and the rigor were only part of the vehicle toward understanding, and this vehicle could easily be overused. There is no real answer to this question. Somewhere there must be a middle ground, but its location is not clearly seen.

The members of Bourbaki retaliated against the criticism by launching their own attacks, in part by pranks they continued to perpetrate on foreign mathematicians. One day, the American Mathematical Society (AMS), headquartered in Providence, Rhode Island, received an application for membership in the Society from "Mr. Nicolas Bourbaki."

J. R. Kline, who was then the secretary of the AMS, knew all about Bourbaki. He sent the application back to France with the following note: "The American Mathematical Society has two classes of membership: individual and institutional. I understand that this is not an application from an individual."[21] "Mr. Bourbaki" was welcome to join the AMS, but "he" would have to pay the (much higher) institutional rate. Ralph P. Boas (1912–1992), who related this story in his article "Bourbaki and Me," published in 1986 in the *Mathematical Intelligencer*, noted that Bourbaki took retaliatory action against the AMS by floating a rumor that Boas was not a real person but rather a collective pseudonym of the editors of *Mathematical Reviews*.[22]

Ralph Boas had served as the executive editor of *Mathematical Reviews* from 1945 to 1950. Because of his position in mathematics, he was asked by the *Encyclopaedia Britannica* to write the annual article on mathematics for the encyclopedia's Book of the Year. In his article for the Book of the Year, Boas mentioned the appearance of several volumes of Bourbaki's work. Since Bourbaki was not known to the general public, Boas decided to explain that Bourbaki was a collective pseudonym of a group of French mathematicians, a fact he had been well aware of ever since meeting André Weil in 1939. At that time, the group was not yet as secretive about its constitution as it would later become.[23]

A short time after the article appeared in the encyclopedia's Book of the Year, Boas received a curious letter datelined: "From my ashram in the Himalayas." It began, "You miserable worm, how dare you say that I do not exist?" It was signed "Nicolas Bourbaki." Boas notes wryly that "here Bourbaki was displaying less than his usual precision of language, since I had not asserted his nonexistence, only his nonindividuality."[24] Bourbaki apparently was not mollified by writing this strong note to Boas, as he then wrote a letter to the *Encyclopaedia*

Britannica as well. The editor, Walter Yust, forwarded this letter to Boas, with a request for clarification. He wanted to know whether or not Bourbaki existed as a mathematician. Boas replied with a short explanatory note about the identity of Bourbaki, and added that Yust might also consult with the well-known American mathematician Saunders MacLane, since both he and Yust were in Chicago.

Yust then wrote a letter to MacLane, and the latter went down the hall at the University of Chicago and showed the letter to André Weil, who was also on the mathematics faculty at Chicago at that time. Weil insisted to MacLane that "he *must* tell the Encyclopaedia that Bourbaki *did* exist." [emphasis in the source][25] MacLane went on to write the encyclopedia's editor a guarded letter that did not say Bourbaki existed, but hinted that he did. This letter, too, came to the attention of Boas in Providence. Boas again wrote to Yust, and even referred him to J. R. Kline, who then told Yust that Bourbaki indeed had attempted to apply for membership in the AMS but had his application rejected on the grounds that he was not a real individual. It was because of Boas's involvement that Bourbaki then spread the rumor that Boas did not exist and was simply a pseudonym. These wars continued for some time.

BOURBAKI HAD A "sister." This was André Weil's sister, the philosopher Simone Weil. André and Simone had grown up together and were very close—as close as a brother and sister can be. While away from her, André would write her long letters, in which he described his deepest work in mathematics. He would preface the mathematical discussions by saying: "You may not understand this, but . . ."

And Simone Weil did try hard to understand mathematics,

and to learn about what her brother's work was all about. She began to come to the Bourbaki congresses, and eventually she became a regular participant in these meetings, whether or not she was a mathematician. As such, she could be described as Bourbaki's sister. Many photographs taken at the early congresses show her standing with the group.

But Simone Weil became a philosopher, and pursued a track in life that was different from her brother's. She forsook her Jewish roots and converted to Catholicism, and her writings exhibit a degree of anti-Semitism, a fact that has been attributed to Simone's self-hate.[26] Her self-hate became more evident when, in 1942, in England, her anorexia nervosa became acute and she starved herself to death. Simone Weil's philosophy was political in nature; she was concerned with the nature of oppression, oppressed people, and oppressive regimes. Her political interests grew into activism when she joined the Republican forces fighting against Franco in the Spanish Civil War.

She has been described as a strange person. People who knew her always commented on the piercing look in her eyes, and the fact that her presence in a room full of people made the others uncomfortable. Although she had an overly intense personality and an overpowering presence, Simone Weil certainly added color to the mathematical meetings of Bourbaki. And since the members of Bourbaki were all male, they certainly appreciated the presence of a woman, especially one who was interested in what they were doing and tried to understand as much mathematics as she could.

Simone Weil produced a voluminous philosophical oeuvre in her short life. In his memoirs, André Weil often writes about his sister's views and opinions, and he draws parallels between his own views and those of his sister. Perhaps André's philosophy of mathematics—and his opinions about the history of ideas and the history of mathematics—had something in common with the philosophical thought of his sister.

FROM THE 1950S to the 1970s, Bourbaki reigned supreme over mathematics. But Bourbaki's greatest contribution was the one it made to Western civilization as a whole, not only to mathematics. That contribution has been the development and promotion of the concept of *structure*. The idea of structure can be mathematical, and, indeed, it is best understood and used within a mathematical framework. But the idea itself was not born in the field of mathematics.

As we shall see, the modern idea of structure originated in linguistics, and its seeds were sown at the turn of the century with the first ideas in modern linguistics proposed by Saussure. These ideas were further developed in the early decades of the century by the Russian linguists Troubetzkoy and Jakobson. André Weil was aware of the emergence of structuralism in linguistics, and his own ideas in this area were crucial to the further development of structuralism by Bourbaki.

Bourbaki took the rather vague notion of structure to a much higher, and more precise, level. Since structure is something that can be very well cast in mathematical terms, Bourbaki had a great opportunity here to make an incredible contribution to civilization: it could formalize, axiomatize, and generalize this concept, making it a very precise mathematical idea. Bourbaki did just that. It did so for its own reasons and purposes: the mathematicians of Bourbaki wanted to use the idea of structure in mathematics. But in making this concept rigorous, they also gave the rest of the world a fabulous tool: the well-defined and precise idea of what structure really means. Within a few years of the mathematical development of the concept of structure by Bourbaki, everyone in science, social science, the humanities, economics, and psychology was talking about structure and using this concept in their works. The age of structuralism had truly arrived once Bourbaki made this concept well-defined.

In his definitive history of structuralism, the French historian François Dosse wrote about the role of Bourbaki in promoting the concept of structure. Bourbaki's mathematical structures brought pure formalism into the context of structures, he noted. "Bourbaki's ideology has certainly strongly contributed to the mentality and activity of structuralism," he wrote. This brought about a new "ideology of rigor" to the concept. "Bourbakism made the edifice of mathematics appear as a splendid construction." This was due, according to Dosse, to the importation into mathematics of the concept of structure, born in the area of linguistics.[27]

"Structuralism and phenomenology were thus directed toward the search for mathematical ideals," Dosse continued, "Those ideals, however, are not the result of evading the real world, nor do they reside outside of the realm of experience. Rather, they are a means of capturing the properties of objects and ideas. They are rooted in a realm of symbolic entities, not relevant directly to the world of the intelligible or the perceptible, but one that occupies some place in between." These mathematical structures were the results of the efforts of the Bourbaki group, which allowed us to construct problematic objects that are only defined through symbols. . . . It allows us to obtain powerful theorems that enable us to control chains of properties of objects seemingly of differentiated nature."[28]

Thus Bourbaki took the concept of structure out of its limited linguistic significance, made it precise and mathematically powerful, and unleashed it on the world of ideas. It would take an anthropologist, however, to then use both the linguistic model and the vastly improved mathematical concept of structure to make structuralism a viable approach in a wide variety of situations in the world.

ten

CLAUDE LÉVI-STRAUSS AND THE BIRTH OF STRUCTURALISM

STRUCTURALISM IS A method of intellectual inquiry that provides a framework for organizing and understanding areas of human study concerned with the production and perception of *meaning*.

Structuralism is interdisciplinary, and multidisciplinary; it has been influencing mathematics as well as philosophy, linguistics, anthropology, and literary criticism.[1]

The emergence of structuralism constituted a revolution in human thinking. Its ideas, which developed in the sciences and in mathematics, quickly took over and invaded other areas of human thought. Structuralism became a dominant philosophy in the West in the middle of the twentieth century. It culminated in France with the wide structuralist movement of the 1960s, a movement that succeeded the preeminent French philosophy of existentialism. The first ideas that were to bring about the structuralist revolution began to germinate, innocently enough, during the interwar period among the members of the Prague Linguistic Circle—of which a key member was the Russian linguist Roman Jakobson. The term *structure* actually began to be used by this circle around

1929.[2] Six years later, it would be used by Bourbaki. It must be noted, however, that many mathematicians—long before Bourbaki—had been looking for structure. Structure, in fact, can best be viewed as a concept that is both amenable to mathematical manipulation and definition, and derives its power from mathematics.

All of mathematics can be viewed as the search for structure—in numbers, spaces, or algebraic systems. Sophus Lie, for example, was interested in "the structure of a group" in the 1880s.[3] But the French anthropologist Claude Lévi-Strauss, rather than any mathematician or linguist, is the person generally credited as being the "father of structuralism." Lévi-Strauss earned this recognition because he was able to take an idea used in linguistics and inherent in much of mathematics and make it useful in anthropology as well as in many other areas of human interest. Lévi-Strauss successfully transformed an academic, or purely intellectual, concept into a useful and very powerful tool of analysis with wide applicability.

While the concept of structure first made its appearance in linguistics, mathematics—especially in the work of Bourbaki—has been the area through which the structuralist intellectual movement of the twentieth century derived its power and meaning. The fundamental assumption of structuralism is that all of human behavior arises from an innate structuring capability. This structuring ability, latent in the human brain, gives rise to language. But the same structures hidden inside the brain also lead to myths, creativity, and various social patterns. Thus the same structures should be identified through study in areas such as psychology and the social sciences. And, in fact, the structures programmed in the human brain do manifest themselves through studies in these areas.

Structuralism deals with the relationships between parts and the whole. Totality takes logical priority over individual parts, and the relationships are more important than the entities they connect. The hidden structure is thus much more important than what is obvious or apparent in any given situation. It is the symbolism that matters, rather than the entities symbolized. Because of the philosophical components of structure, we find the ideas of structure in areas far beyond the sciences. For example, the philosophers Jacques Derrida and Michel Foucault are structuralists, as were the linguists Roman Jakobson and Nicolai Troubetzkoy, whose ideas helped found this discipline, and the psychologists Jacques Lacan and Jean Piaget, among others.

CLAUDE LÉVI-STRAUSS WAS born in Brussels to French parents on November 28, 1908. His was an intellectual family that included among its members painters and violinists. Art and nature were the two most influential elements in the boy's life. In the countryside, where his parents had bought a house, to his great delight, he would spend many hours hiking, often as long as ten or fifteen hours in a day. With a mind that was placed somewhere between art and practicality, Lévi-Strauss sought to impose logic on the esthetic elements in his environment.[4] It was this search for logic, even in areas governed by feeling or intuition, which would bring him to the idea of structuralism.

Since his days in high school, Claude Lévi-Strauss was interested in social issues, including politics, and this interest would lead him to study the nature of human society. Anthropology was in its infancy during this period and was not yet the science we view it today. Lévi-Strauss would turn anthropology into a modern science. His involvement with social and political issues while he was still a student got him elected, in 1928, general secretary of the Federation of Socialist Students in France. He also worked as secretary to a socialist French deputy. But soon Lévi-Strauss would need to abandon his political work so he could study hard in preparation for his degree in philosophy, which he was awarded in 1931, third in his class.

His career as an ethnologist was launched, as he recounted in his masterpiece *Tristes Tropiques* (published in 1955), one Sunday in the fall of 1934, when he received a telephone call from Célestin Bouglé, the director of the École Normale Supérieure in Paris. Bouglé suggested to Lévi-Strauss that he agree to have his name placed on the list of candidates for a professorship in sociology at the University of São Paulo in Brazil. (As Lévi-Strauss recalled it, Bouglé naively thought that the suburbs of the Brazilian metropolis were teeming with native Indians, and, once Lévi-Strauss was selected for the position, Bouglé suggested enthusiastically to the successful candidate that he spend his weekends in these suburbs studying native societies.)[5] Lévi-Strauss jumped at the opportunity, not because he sought adventure or liked to travel, but because he was eager to get away from speculative philosophy and plunge head-on into the nascent science of anthropology.

Doing field work in Brazil in the 1940s, Lévi-Strauss found that he had great difficulty classifying Amerindian languages he encountered. He became very interested in linguistics, believing that this field held a special place among the social sciences because of its precise methods and analyses. He sought out linguists who might help him solve the problems he was facing, but it was not until some years later, in New

York, that he would finally meet a linguist who could give him that help.

Lévi-Strauss returned home to France in 1939. But he arrived there just in time to have to leave again: World War II started, and he had to escape the advancing Nazis since he was Jewish. He was fortunate enough to be invited to assume a professorship at the New School for Social Research in New York. Lévi-Strauss boarded the *Capitaine Paul Lemerle*, headed for America. Among his fellow passengers on this ship fleeing Europe were a number of European intellectuals also on the run.

As soon as Claude Lévi-Strauss arrived at the New School in New York, it was suggested to him that he immediately change his name. He was told that if he didn't change his name, everyone he met in America would assume he was selling blue jeans. While in America, therefore, Claude Lévi-Strauss was known as Claude L. Strauss. This did not prevent confusions, however, and he later recounted that never a year passed in which he didn't receive at least one order for a pair of jeans, "usually from somewhere in Africa."[6]

At the New School, Lévi-Strauss met the Russian linguist Roman Jakobson, who had begun to infuse structural ideas into phonology, following up on earlier ideas developed by another Russian linguist, Count Nicolai Troubetzkoy. Jakobson was testing out these new ideas in courses he was teaching on structural phonology at the New School, courses he conducted in French. Lévi-Strauss sat in on Jakobson's experimental classes, and found them very illuminating. Jakobson, in turn, audited Lévi-Strauss's own courses at the New School on human kinship. The two men hit it off well and became close friends. Their friendship would last a lifetime, and through the cooperation between these two giants in their respective fields structuralism was born.

Jakobson was very impressed with Lévi-Strauss's doctoral dissertation, which the latter had shown him, and Jakobson strongly urged his friend to start editing his thesis in preparation for publication. Indeed, this seminal thesis would be published in 1943 under the title *Elementary Structures of Kinship*. It would become one of the most important books ever written in anthropology, and one of the classics in all of science.

But as Lévi-Strauss was preparing his thesis for publication, he came to the realization that his work was incomplete. The thesis was a study of kinship systems in Australian aboriginal tribes, and it attempted to explain why people of certain groups within certain tribes were allowed to marry only certain people and forbidden from marrying others. The research paper was thus an empirical or descriptive analysis of who was encouraged to marry whom, who was forbidden from marrying whom, and so on. But the more he thought about the problem, Lévi-Strauss realized that there must be something more to the problem—something that was missing from his analysis. By then he had been exposed to Jakobson's ideas about structure, and he knew that his new friend had identified in linguistics the essential elements of phonology—the hidden basics of all language. Could a similar analysis be useful here, in anthropology? he wondered.

In describing his thinking about the lessons that anthropology in general and his project of understanding kinship in particular could learn from linguistics, Lévi-Strauss wrote the following.[7]

> The illustrious master of phonology, N. Troubetzkoy, furnishes us with the answer to this question. In an article, he explains the phonological method in four fundamental levels. In the first place, phonology passes from the study of *conscious* phonological phenomena to their *unconscious* infrastructure. One refuses to treat the *terms* as independent entities, but rather taking them as the basis of an analysis of the *relations* that exist among these terms. One then introduces the notion of a *system*. The actual phonology then consists of the structures that are thus revealed. Deducing these structures logically endows them with an absolute character

But the initial attempt to apply the principles of structural analysis, which Troubetzkoy and Jakobson had employed so successfully in linguistics, to the field of anthropology produced great difficulties for Lévi-Strauss. He described these problems as follows:[8]

> However, one preliminary difficulty on the road to applying the phonological method to the study of primitive societies strikes us immediately. The superficial analogy between phonological systems and kinship systems leads in a false direction. It consists of assimilating, from the point of view of their formal treatment, the terms of kinship within the system of phonemes of language. We know that in order to arrive at the law of structure, the linguist analyzes the phonemes as "differential elements," which one can then organize as "coupled opposites." The sociologist is then tempted to try to break down the terms of kinship into a similar system, using an analogous method.

But the structures of language, which revealed themselves to the linguists who carried out the analysis, were apparently of a different nature than those that Lévi-Strauss faced in his kinship systems analysis. In other words, the linguists were able to "break the code" of language using a nonmathematical investigation. Lévi-Strauss's system of human kinship was much more complicated. In using the analogy of structural linguistics, he had indeed come to the right idea. But the tools he needed in order to achieve his goal would prove to be more complex than those used in linguistics.

Lévi-Strauss knew that the problem he was studying had a deep underlying structure. And he understood as well that something along the lines of the structuralist approach used in linguistics by Troubetzkoy and Jakobson should be used to attack his problem. But he couldn't do it. His problem required a much deeper, mathematical analysis, for sometimes the codes underlying a problem are simple enough to break without mathematics; other times, the codes—the structural foundations of a problem—are so complex, that they require the tools of a mathematician in order to uncover them.

Lévi-Strauss was very fortunate—for he would soon make direct contact with Bourbaki. In New York, he would meet Bourbaki's cofounder, André Weil, and the latter would use group theory to formally break the code of human kinship and expose the internal structure of this important problem in anthropology as the structure of a certain mathematical group.

Once the solution was obtained, and Lévi-Strauss, with the help of Weil, solved the problem of kinship, the publication of the resulting book by Lévi-Strauss—with a mathematical appendix authored by Weil—would be widely viewed as the "act of creation of structuralism."

IN HIS WORK on kinship systems, Lévi-Strauss was looking for the *inner structure* underlying his observations. His first ideas in this area occurred to him long before he came to New York and met Jakobson and later Weil. Lévi-Strauss's first ideas about internal structure in anthropology occurred to him during his stay in Brazil. These first notions of structure became more pronounced in his mind as he started to study the data for Australian aboriginal societies. Here he could see both similarities with the observations he collected in Brazil and results that were contrary to the Brazilian data. The comparison of information about primitive societies that were separated by thousands of miles of ocean made Lévi-Strauss want to seek the constants that underlay these very different societies.

In particular, incest was forbidden in all societies that he and other anthropologists studied, and Lévi-Strauss was searching for the underlying structure of the kinds of incest that were forbidden. The universality of the prohibition of incest became the object of study for Lévi-Strauss. He was looking for the hidden internal structure of kinship in all societies, and he began to use the structural approach. Instead of concentrating on the negative side of the prohibition against incest, Lévi-Strauss started to look at it from a positive direction. The prohibition against incest seemed to play a positive social role in addition to its biological function of preventing inbreeding: it made it possible for members of a society to form alliances. Approaching the problem from this positive, rather than negative, direction brought Lévi-Strauss to his structuralist breakthrough. Using his method of reasoning, he was able to demonstrate that a set of rules governing who may marry whom in a given native society allowed members of the society to form important and useful alliances. And this analysis brought him to the beginning of an understanding of how kinship works in native societies.

Ideas about structure, imported from linguistics, and the work in that field by Troubetzkoy and Jakobson, afforded Lévi-Strauss early limited success in transforming anthropology from a physical-biological social science without a deep underlying scientific basis into a modern discipline within which the information about various societies and their kinship systems could be analyzed in a systematic way that could reveal the internal structure of the system. But he had not broken the code yet. Structuralist thinking was working in principle—but concrete results eluded him, for he was in need of help from a mathematician. The hidden structures of kinship were too complicated to be revealed without detailed mathematical analysis.

Eventually, once the social code of kinship systems was broken, the new structuralist approach would reveal the salient points about a society under study and expose the latent universal rules that govern human behavior.

IN 1943, CLAUDE Lévi-Strauss met André Weil in New York, and the exchange of ideas between them was what eventually led to Bourbaki's ideas being introduced into anthropology. The first important example of the use of mathematical principles in anthropology was to be the solution of the difficult marriage-rules problem in tribes of Australian aborigines studied by Lévi-Strauss. This problem was solved by André Weil using purely abstract algebraic methods. Weil was so proud of his mathematical solution of Lévi-Strauss's problem, and the connections forged between "pure" mathematics and applied science, that he continued to tell the story about this cooperation between practitioners of different fields until his death.

Lévi-Strauss had large amounts of data, but his initial analysis had completely failed to result in the desired rule he tried to extract from the data. It seemed that everything he had tried failed, and so he came to the inescapable conclusion that only a mathematician could help him solve the mystery. Thus he went to see Jacques Hadamard, the famous French mathematician and uncle of Laurent Schwartz, who was also living in New York at the time, one of a group of European intellectuals who had fled Hitler and made up the émigré community that has been called "Paris on the Hudson." Hadamard, who was by then quite old, did not help Lévi-Strauss. He told him: "Mathematics has four operations, and marriage is not one of them."[9]

So Lévi-Strauss went to see André Weil, who was in New York at the time. Weil's reaction was quite different. Weil was a superb algebraist, and while his work had always been in pure mathematics, he quickly saw that Lévi-Strauss's problem could be attacked using the abstract theory of groups, here applied to a real-world problem. In fact, Weil was famous for saying, whenever anyone was stuck on a mathematical problem, "When in doubt, look for the group!" Weil worked on the problem for a while, and solved it. What was a very complicated situation in the real world was solved brilliantly by an application of abstract algebraic techniques.

Weil explained to Lévi-Strauss that he was able to solve the problem by ignoring the actual elements of the problem: the nature of the marriages themselves. Instead, Weil concentrated on the *relationships* among the marriages.

This idea reflected the main thought of Bourbaki: that relationships and structures were the key elements of mathematics.

The solution of his problem by the Bourbaki cofounder made Lévi-Strauss even more interested in structures and structuralism, and this encounter, along with his interactions with Jakobson, made him concentrate all his efforts on developing structural methods in anthropology. Eventually, this work would bring structuralism to prominence.

Lévi-Strauss's data on marriage rules indicated that among the Australian tribes he studied, there were four population groups, which, for simplicity, he denoted by A, B, C, and D. The rule that seemed to exist in this aboriginal population dictated that a man from group A could marry a woman from group B; similarly, a man from group B could marry a woman from C; a man from C a woman from D; and a man from group D could marry a woman from group A. Pictorially, the marriage rules are as shown below.

$$A \to B \to C \to D \to A$$

The question was what happened long-term in such a population. What did these rules say about the development of long-term kinship ties within this society? What was the hidden internal structure of the complex relationships among the members of this society? This was what mathematics was called on to solve in this context.

What Weil did was to define the possibilities of marriage within the society. There were, he understood, four possible kinds of marriages within this aboriginal population. Weil defined the types of marriages as follows:

M1=[man from A, woman from B]
M2=[man from B, woman from C]
M3=[man from C, woman from D]
M4=[man from D, woman from A]

Then, Weil looked at the definitions of the *offspring* of these marriages, and *their* possible marriages. He conducted the analysis, and identified a mathematical *group* that governed this particular process. Mathematics allowed him to solve this problem. Then he wrote up his results as an appendix to the first part of Lévi-Strauss's book, *Elementary Structures of Kinship (Les Structures Élémentaires de la Parenté)*. (The Hague: Mouton, 1947).

In the appendix, Weil wrote, continuing the description of the types of marriages allowed in the Australian aboriginal society that Lévi-Strauss had studied:[10]

> We add to the above that the children of a mother of class A , B, C, D are respectively defined as members of class B, C, D, A. This gives us the following table:

Type of marriage of parents:	M1	M2	M3	M4
Type of marriage of son:	M3	M4	M1	M2
Type of marriage of daughter:	M2	M3	M4	M1

Weil pointed out that what we have here are *permutations* of the marriage types of the parents. What permutation means, he explained, was that the orders of the kinds of marriages in the second and third rows are simply *rearrangements* of the symbols in the first row.

Permutations are studied by the abstract algebra theory

of groups. As we have seen earlier, permutation groups are a common example of a mathematical group. Recall that the permutation group of three elements consists of the members ABC, ACB, BAC, BCA, CAB, and CBA. Weil then applied the powerful results of the mathematical theory of groups, in particular the study of permutation groups, to the analysis of marriage types. He pointed out that results from group theory allow us to make conclusions about such a population; in particular, the kind of group under analysis here could be *reducible* or *irreducible*. A reducible group of marriage types would indicate to the anthropologist that the population under study is really the grouping together of two distinct populations that never intermarry and perhaps simply occupy the same region. If the group inherent in the permutations above is irreducible, however, then this means that the population is unified and cannot be naturally broken into two or more subpopulations that do not intermarry, as happens in this example.

Each of the populations studied by Lévi-Strauss had its own mathematical group structure. Once the structure was analyzed, it was possible to determine whether or not the population had noninteracting segments, or whether, on the other hand, it was uniform. This was a determination that could come about only through mathematical analysis. One could study the population for years and not find this important characteristic if one did not employ the structural mathematical analysis using group theory. For what Weil discovered was tantamount to a theorem: it was something that had to be true about a population because of its inherent structure and nothing else. Simply not finding any interaction between subgroups, if performing a strictly empirical analysis, would never be enough because one would never know whether a connection between the subgroups was *possible* (in which case, in the future it *could* appear) or whether it was not possible. Only mathematics could give the definitive answer.

Weil's analysis was *structural,* and it was in the spirit of Bourbaki and its aims. Through the results derived from purely abstract mathematics, Weil was able to discover an inherent truth about the populations under study. This truth could not have been discovered by the weaker, non-mathematical structural methods that had been used by the linguists. The power of mathematical structural methods had thus been brilliantly demonstrated. Lévi-Strauss was duly impressed—and grateful—and from then on dedicated all his efforts to promoting the use of rigorous structural methods in anthropology and other areas. Lévi-Strauss took Weil's ideas and learned from them. His entire book about kinship—a classic of anthropology and of structuralism—is unique in that it infused into anthropology the structural ideas of the new mathematics of Bourbaki.

In time, Lévi-Strauss's book became the opus magnum of French structuralism. In it, he tried to explain the systems of kinship and marriage—which come in an enormous variety throughout societies around the world—by means of a single principle: the principle of exchange. Thus the marriages themselves do not matter at all. What matters, rather, are the relationships among the various kinds of marriages, as Weil had been able to demonstrate using the results of the abstract theory of groups. This idea, in fact, is the heart of structuralism. Structuralism teaches us that the elements under study are unimportant—only the relationships among these elements have significance. Thus, in the example Weil analyzed, the pertinent variable was how marriages of the parents affected the kinds of marriages their children could have.

The *exchange* that Lévi-Strauss studied is perceived to be a manifestation of fundamental structural constants within the human mind. And these structural constants may be found in other systems of human culture. One example of such a manifestation of hidden structure is language. It was for this reason that the influence of structuralism in anthropology originally came from the work in linguistics. The same structural ideas manifest themselves in psychology and other fields.

THE GROWING SUCCESS of the notions of system and structure found unexpected usefulness within many diverse areas and disciplines during the twentieth century. Notably, structuralism demonstrated a great capacity to explain the interdependencies of the elements constituting a discipline. This trend was obvious in linguistics, anthropology, sociology, biology, and economics.[11]

Structuralism successfully replaced the earlier philosophy of science: empiricism. The idea of structuralism is that empirical information—descriptive analysis of data—is not enough, and does not explain the underlying phenomenon inherent in the data. Structuralism looks for the hidden structure within the data, rather than merely describing the way the data appear.

Anthropology had been part of the natural sciences until the point at which Lévi-Strauss made his breakthrough discovery of the structure of human kinship systems. Until then, anthropologists simply collected data and studied their observations using the empirical method. But Lévi-Strauss's genius was that he saw that a deeper approach could lead to amazingly powerful results.

EXISTENTIALISM, WHICH FLOURISHED in France in the years following the liberation of the country by Allied Forces in 1944, signaled a dawn of intellectual activity. The existentialists believed in a philosophy of *engagement*, which set them apart in those early troubled years from science and its potential evils, prime among them the bomb. Of all the sciences, existentialists approved only of psychology, and in particular they hated mathematics. While existentialism monopolized French intellectual debates during this period, other intellectual directions were being explored. It was in this context of new ideas that Claude Lévi-Strauss's book made its debut: this monumental intellectual masterpiece was the "act of foundation" of structuralism, the new cultural direction in France and elsewhere.[12] In addition, François Le Lionnais, a mathematician and writer, published in 1948 a special issue of the journal *Les Cahiers du Sud*, in which he delineated what he called "Great currents in mathematical thought." This issue included a key programmatic statement about the work of Bourbaki. Le Lionnais articulated his overall approach to a unified science of mathematics along the lines of Bourbaki.[13]

These twin publications set the stage for the modern development of ideas in the second half of the twentieth century: Lévi-Strauss's book would lead the way in the social sciences, and Bourbaki would do so in mathematics. Both would promote the ideas of structuralism. Structuralism was a deep method that stripped away all the unnecessary elements of a system, leaving only the most salient points of the theory. Lévi-Strauss thus did to anthropology what Jakobson did to linguistics—and what Bourbaki brought into mathematics. Bourbaki strove to center all of mathematics on the important elements of structure:

> In conclusion, we can reasonably say that the intersection of Lévi-Strauss, Jakobson, and Weil, in New York in 1943, by cross-breeding anthropology, linguistics, and mathematics, helped make structuralism possible. And although the dialogue between mathematics and structuralism failed to be sustained, this fortuitous encounter was the seed of a lasting cultural connection.[14]

At first, Lévi-Strauss's *Elementary Structures of Kinship* was embraced by the existentialists. Simone de Beauvoir reviewed the book very favorably in *Temps modernes* in 1949 (vol. 5, 1949, pp. 943–9). But only in the late 1950s did the book begin to exert real impact on French intellectual life. By 1968, structuralism as championed in this book would replace existentialism at the apex of French philosophy. Lévi-Strauss and others such as Michel Foucault would then take Sartre's place at the center of modern philosophy in France.[15] In addition to Lévi-Strauss and anthropology, Roland Barthes used structuralism in literary and cultural criticism, Jacques Lacan used it in psychoanalysis, and Michel Foucault used it in philosophy. Other fields enjoyed the fruits of this revolution as well. In 1951, Lacan, Lévi-Strauss, and Benveniste began meeting regularly with the mathematician Georges-Théodule Guilbaud in order to work on structures and attempt to uncover links between the social sciences and mathematics. The ideas germinated, and from the late 1950s until the end of the 1960s—the period in which Bourbaki made its greatest contributions to mathematics—structuralism "happened."[16]

Structuralism, with its mathematical underpinnings, became a major social and cultural phenomenon. The trend became important after the publication of yet another book by Claude Lévi-Strauss, *Structural Anthropology*, published in 1958 (New York: Basic Books).

Following the publication of this book, two conferences were held in 1959 with the purpose of exploring and explaining the meaning of structuralism. Notions of mathematical structure developed by Bourbaki figured prominently in these conferences.

Thus mathematical structures became the key idea in these congresses devoted to the social sciences and the application of structures in these areas. The first of these two meetings was held in Paris January 10–12, 1959, and it was named "Meanings and Uses of the Term Structure in the Human and Social Sciences."

The second congress, held between July 25 and August 3 of the same year, took place at Cerisy-la-Salle, and its theme was "Genesis and Structure." The key players in world structuralist movements were there: Lévi-Strauss in the Paris conference, and Jean Piaget in the meeting at Cerisy-la-Salle. Mathematics exerted a universal appeal in these congresses and in ones that would follow in the years to come.

> Bourbaki had seeped into intellectual folklore because of his high profile in the mathematical community and his alleged role in educational reforms. He had become a synonym for rigor, axiomatics, and set theory.[17]

Many authors, however, kept insisting that structuralism could be pursued without mathematics.

From the structuralist work Jakobson had done in linguistics, Lévi-Strauss learned not to get bogged down by the multiplicity of terms, but to look for the simplest and most salient relationships uniting them. Jakobson looked for the smallest unit of spoken language: the phonemes. Lévi-Strauss, similarly, looked for elementary structures in anthropology. Both Lévi-Strauss's structural analysis of kinship and Bourbaki's structural view of mathematics aimed at unifying their respective disciplines by concentrating on underlying structures. Structuralism and Bourbakism peaked in the late 1960s, and then began a decline as postmodernism took over.

IN 1948, LÉVI-STRAUSS returned to France and held a number of temporary positions. He was head of research at the CNRS, and later deputy director of the Museum of Man in Paris. He was then elected at the École Practique des Hautes Études, to the chaired professorship of Religions of Uncivilized Peoples.

At some point after this appointment, a number of African students asked to meet the new chair. "Can you say that people who come to discuss ideas with you here at the Sorbonne are uncivilized?" they asked the professor. The name of the area of study was quickly changed to Religions of Peoples without Writing.[18] In 1959, Lévi-Strauss was elected to a chaired position at the prestigious Collège de France.

eleven

THE LINGUISTIC ORIGINS

WHILE STRUCTURALISM WAS instigated by the work of Bourbaki, its roots—as evidenced in the work of Claude Lévi-Strauss—had nonmathematical aspects as well. As we have seen, the germ of the idea of structuralism, stripped of its explicitly mathematical aspects, was developed by the Russian linguists Nicolai Troubetzkoy and Roman Jakobson.

ROMAN JAKOBSON WAS born in Moscow on October 11, 1896. As a child, he began to read voraciously at the age of six and devoured many books on a variety of topics. Language held a special interest for the child, and early on he learned on his own to speak French and German. Then he turned his attention to poetry and by the age of twelve had read all the great Russian poets as well as many foreign writers. And, similarly to his future friend and colleague, Claude Lévi-Strauss, Jakobson was very interested in painting. To him, painting was the pinnacle of creative culture.

In 1915, Jakobson was instrumental in the foundation of the Moscow Circle of linguistics, which was devoted to the promotion of linguistics and poetry. The first meeting of the Moscow Circle took place in Jakobson's parents' dining room. But the meeting of intellectuals in tsarist Russia was a very dangerous activity, and soon the Circle chose even more hidden venues for its meetings.

In the meantime, Jakobson made the acquaintance of Prince Nikolai Troubetzkoy, a major force in linguistics. Troubetzkoy introduced Jakobson to new ideas and told him about new work in phonology that had been done in France by the school of Antoine Meillet, work that was reviving linguistics from its stagnation to a field worthy of study, thanks to new developments in phonology. Jakobson moved to Saint Petersburg, and on the eve of revolution in 1917 helped found the Saint Petersburg Circle of linguistics—a group dedicated to the study of poetry from a linguistic point of view. This group included a number of Russian poets such as Eichenbaum, Polivanov, and Yakoubinski.

Like Grothendieck, who became attracted to poetry when he was growing up in the camps, and like the Oulipo group of writers and mathematicians who would follow him in the 1960s, Jakobson was enthralled by the power of poetry and captivated by the hidden structures of language he saw in poems. In poetry, Jakobson clearly saw a coherence that one rarely discerns as easily in other parts of literature, or in speech. In poetry, the total was much more than the sum of its elements; there was clearly something more. And this something extra he found in poetry was the inherent structure. Linguistics thus placed Jakobson right in the middle between creation and science. But such artistic and scientific beauty could not flourish in revolutionary Russia, and Jakobson decided that political upheaval was not to his liking. So he moved to Prague.[1]

Jakobson arrived in Prague as an interpreter working for

the Russian Red Cross. Later, he began translating Russian poetry into Czech, and Czech poetry into Russian. As soon as he began to work seriously in this area, he discovered that differences in tonality were the most important ones between the poetry of the two languages. What worked in his favor and made his discovery possible was the fact that the two languages share common roots and are similar in other ways, but the tonal differences, most clearly perceptible in poetry, made for a significant difference between the two languages.

Structural phonology was thus born through the interactions among natural language, cultural language, and poetic language. In Prague, Jakobson again met with Prince Nicolai Troubetzkoy, whom he had known since 1915. The latter had fled the Russian revolutionaries, who in 1917 would have put him to death, and took refuge in Prague. There, on October 16, 1926, Jakobson, Troubetzkoy, and a number of Czech poets and linguists founded the linguistic Circle of Prague.

The Circle of Prague set as its agenda the study of linguistics, with structuralism at its core. In 1929 the Circle of Prague began to publish its proceedings, "The Theses of 1929," with the aim of setting the agenda for the study of linguistics for generations to come. The structuralist idea was to make a distinction between internal language and actual language. The Prague linguists wrote:[2]

> In its social role, one must distinguish language following the existing relationship between it and the extra-linguistic reality. Language has a function in communication, that is, it is aimed at the signified; or it has a poetic function, that is, it is aimed at the symbolism itself.

The stated purpose of the Circle of Prague was the study of the second meaning of language, its role in poetry, that is, its relevance to the symbolism itself—rather than the literal meaning signified by language.

In order to support himself, Jakobson took up a position of professor of linguistics at the University of Brno—a post he would hold until the advent of the Nazis in 1939 would force him to flee. But Jakobson spent all his extracurricular time working with the Circle of Prague on the structuralist approach to linguistics and was elected vice president of the Circle.

Structuralism in linguistics made its debut on the world stage during the International Congress of General Linguistics, held in The Hague, Holland, from April 10 to 15, 1928. The term "structural linguistics" was first used in this Congress when, during the first two days of meetings, members of the Circle of Prague explained their ideas about structure in linguistics and the distinction between symbols and the elements they signified, as well as the role these took both in language as a form of communication and language used in poetry. In the *Archives of the XXth Century* Jakobson described the presentations made by the Circle of Prague at this congress: "We have posed the question of structure as the central question, without which nothing in linguistics may be treated."[3]

Jakobson had strong connections, also, with the linguistic Circle of Copenhagen, which was similarly devoted to new ideas in linguistics and published the journal *Acta Linguistica*. When the Nazis occupied Czechoslovakia in 1939, Jakobson escaped from Prague and took refuge in Copenhagen. There he continued his linguistics work, now within the Circle of Copenhagen. But soon, as the Nazis continued the blitz through Europe, he had to flee again. He crossed the Baltic Sea and sailed to Norway, but as the Nazis further advanced he crossed the border to Sweden. Even there he felt unsafe, and in 1941 he arrived in New York. Here he made contact with American linguists, and the cooperation with them brought him new ideas. He explained these ideas in his courses on structural linguistics at the New School for Social Research, where he made the acquaintance, and the friendship, of Claude Lévi-Strauss.

Jakobson and his ideas were welcomed in New York. The journal founded by American linguists, *Word*, listed him among its members of the editorial board. The first issue of the journal, appearing in 1945, was devoted to structural concepts in linguistics. It was also concerned with cooperative projects between American and European linguists. Structural phonology was seen as the "model of all models," the rational kernel of structuralism. The groundwork had been laid out in Prague, when Troubetzkoy wrote his *Principles of Phonology* in 1939, in which he defined phonemes by the place they occupied in the phonological system. His system consisted of relating phonic opposites to one another, taking into consideration four distinct traits that had been identified: Nasality, Point of articulation, Labialization, Aperture

Troubetzkoy then looked for pertinent differences; he wanted to find the minimal units of pertinence, and here he found the phoneme. The key idea was now to search for the internal laws that determine the code of the language. Jakobson continued to work on these ideas developed in Prague, and defined a table in which he listed all the pertinent traits of the language based on these elements and a dozen binary pairs of opposites.

He assumed that the elements in this table accounted for

all the opposites that could be found in all the languages of the world. This was the structuralist's dream, to study the internal structure of language and through the analysis to identify the universal invariants of language—the phonetic rules that govern every possible language. These invariants underlie the variability that one sees in the data. Thus, what the structuralist scientist does is to look behind the variability of the data to find the real factors affecting the situation in question.

On a level that was equal to the mathematical formalism of structural analysis—the approach taken by Bourbaki—Jakobson also looked for the phonetic binary code. This binary element is evident in language from a person's infancy, and it lies at the heart of the phonologic system. There is a dualism underlying the data, the dualism between the symbol and the element it signifies. This was the structural essence of language. Phonology is the element that carries the structure of the language. Jakobson arrived at this great discovery having studied the basis conjectured by Troubetzkoy. Once he found it, he realized that the structural approach he pioneered should be applicable in other areas of human study, particularly in psychology and psychoanalysis, as well as anthropology.

The elements of structure are embedded in the human brain and determine its activity in a mathematical way. Thus, what was found in language should appear in other areas as well. Studying Freud, who was one of his intellectual heroes, Jakobson realized that analysis of speech problems could reveal structural problems of the human mind studied by psychoanalysis.

THESE FIELDS, THEREFORE, psychology, anthropology, linguistics, sociology, and economics (which studies the aggregate economic behavior of people) all involve the structural elements that are hidden inside the human brain and the human subconscious. To reach that subconscious level, one must study the behaviors of individuals and societies and seek to identify the latent structures in the brain.

The way the brain processes information, according to Lévi-Strauss, is by using symbolism. The symbolism is what structural analysis is designed to uncover. Structure is thus a code, consisting of concise symbols. The symbolism inherent in brain function follows mathematical rules that are tantamount to the ideas developed by Bourbaki: the notions of closeness, transformation, groupings, and other of the "mother structures" studied by the group.

The symbols in the brain determine how information is processed, but they also capture the workings of language, as uncovered by structural linguists. These codes determine how societies behave—and it was here that Lévi-Strauss made his seminal discovery, for, using the structuralist approach to his discipline, he was able to demonstrate something that would not have been possible otherwise, namely that the prohibition against incest has a positive role in the development of social structures. The discovery of structure is the discovery of codes, how codes develop, and how codes change from one to another.[4]

Proceeding into other areas, Lévi-Strauss set the idea that the code precedes the message—be it in anthropology, psychology, sociology, or linguistics. This is the canonical message of structuralism. Structuralism aims at studying the code as such, rather than concerning itself with the actual message in the code and its context. This is in line with the work of

Bourbaki in mathematics, in which the elements themselves have no meaning; only the relations among them, the codes that exist within mathematics, are of importance.

Thus, according to Lévi-Strauss, structuralism goes beyond the confines of any one discipline. It is an immensely powerful mathematical way of unlocking the inner workings of the brain or of any logical system, and it therefore finds applications in almost any discipline. Structuralism certainly took the social sciences out of their empirical stagnation and brought them into the scientific milieu of the "hard sciences." Thus, Jakobson has said that structural linguistics is "like quantum mechanics."[5]

"We want to learn from the linguists the secret of their success," Lévi-Strauss wrote. "Couldn't we, too, apply to a complex field of our studies the rigorous methods whose usefulness linguistics verifies every day?"[6] Lévi-Strauss also made the important deduction that the symbols—the basic elements in language as well as in anthropology—are mathematical entities embedded in the human mind. Access to the subconscious, in which these elements reside, passes through language, as well as through social behavior. This realization unified, from a structural point of view, the three fields of psychology, anthropology, and linguistics. The unifying elements were the mathematical symbols in the subconscious brain. Thus, structural mathematical analysis could reveal the inner workings of the mind and language and the behaviors of societies. Symbolism reigns in this realm, since the symbols are the latent structures in all three fields, and the symbol thus is more significant than what it signifies, be it in the mind, language, or social behavior. Thus the symbols that structural analysis reveals are the essential elements of the theory.

The next stage in linguistic structuralism brought

structuralism into literature through the work of the French intellectual Roland Barthes. In 1953, Barthes published his book, *Le Degré zero de l'écriture (Degree Zero of Writing)*, which brought new structuralist perspective to literature. His book aimed at identifying a formal reality that is independent of language or style. He thus took the structuralist idea to a higher and more general level.

Barthes began his literary career as a follower of Sartre, and adhered to the Sartrian notion that the act of writing liberates the soul. But he brought this idea into a structuralist setting by separating writing itself from its contents. Language thus becomes the key element, rather than what the language signifies. In the sense of Lévi-Strauss, Barthes looks for the symbols of language, rather than what these symbols signify. In the sense of Bourbaki, the inherent mathematical symbols—regardless of their context—are the most important aspect.

Thus, "point zero" of literature is its symbolism for symbolism's sake. Barthes' idea is thus analogous to that of Lévi-Strauss, who searched for "point zero" of kinship, or Jakobson, who sought "point zero" of language.[7] The "point zero" in literature is, in fact, the contract between the writer and society; it is the basic structural element of literature. Barthes' work brought a new style into literature, a new form of fiction writing, one that was outside the usual norms of writing.

Roland Barthes was born in Cherbourg, France, in 1915. His father died in World War I before Roland was a year old. The child was raised by the mother, and she tried to give him as much affection and attention as possible to make up for the loss of a father. Barthes moved to Paris and attended one of the best high schools in France, and in 1935 he was accepted to study classics at the Sorbonne. While a student there, he cofounded a theater group which performed innovative renditions of classical Greek plays in Paris. But soon Barthes came down with a crippling pulmonary illness, which prevented him from graduating and thus ruined his chances for a formal academic career. He traveled widely, however, and produced literary works. For several years, he submitted a literary article to the French magazine *Les Lettres Nouvelles* once every month.

His voluminous writings earned him a professorship at the Collège de France in 1976, despite his lack of academic credentials. In this respect, he was similar in his life's trajectory to a number of structuralists who also had to establish themselves on their own rather than within the usual academic track in France.[8]

Barthes began to study the popular culture and myths. He sought to "break the masks" and expose the actual elements governing culture, separating reality from myth. His study of popular culture and its mythology brought him to the notion of symbolism, and thus to structuralist thinking. Studying the differences between language as symbolism and language as conveyor of information, Barthes came to linguistic analysis and to the formal structuralism in linguistics. He reached the conclusion that myth separates out reality and exposes the latent symbolism or structure of the language. Using these ideas, Barthes was able to use structural linguistics to infuse into literature the ideas of an inherent mathematical-like symbolism.

During the 1950s, the early days of the structural revolution, Barthes actively participated in a theatrical review, titled *Théâtre populaire*. Through his writings, he contributed to the infusion of structural ideas into theater. He was also involved in a production of Berthold Brecht's *Mother Courage* in Paris. In

Brecht's works, Barthes saw a structuralist philosophy applied in theater; he viewed Brecht as successfully implementing in theater the same ideas he himself was attempting to use in literature and linguistics. Analyzing *Mother Courage*, Barthes came to the realization that Brecht placed in his play the symbolism above the reality. Brecht's plays, therefore, were a manifestation of structuralist ideas in theater.

Barthes' philosophy was centered on symbolism and the role of the symbol in revealing latent structure. This view brought new depth to the work in linguistics that came after structuralist ideas had seeped into other areas, and it widened and deepened the structuralist revolution.[9]

Mythologies and their role as separators of symbolism from reality again found resonance in the area of psychology, and thus, through the work of Barthes, literature joined linguistics, anthropology, psychology, and mathematics in the new structural revolution that swept European intellectual circles in the middle of the twentieth century. But writers found another way into structuralism—directly from mathematics with its elements of combinatorics, topology, and structure. These writers would form their own Bourbaki-like group and explore these new structural ideas.

twelve

STRUCTURALISM IN PSYCHOLOGY, PSYCHIATRY, AND ECONOMICS

WITH THE INFLUENCE of Bourbaki, structuralism entered psychology. The Swiss psychologist Jean Piaget, in particular, was interested in using mathematics in psychology, and here the ideas of Bourbaki were especially useful to him.

"A critical account of structuralism," Piaget wrote in 1968, "must begin with a consideration of mathematical structures."[1] The structural laws that Bourbaki introduced, and which eventually found their way into the sciences, the humanities, and other areas, are:

Composition
Neighborhood
Order
Equivalence

Based on several decades of research in psychology, Piaget concluded that at every stage of the development of intelligence in children, thought processes occur in very structured ways based on the mathematical ideas above, the same ones now made precise by Bourbaki. Bourbaki's structures were crucial for Piaget's work on intelligence. Piaget believed that the acquisition of propositional logic was the key element in the intellectual maturity of a child. The mental structures that enable teenagers, and adults, to think logically are themselves modeled on mathematical structures. The structure of a group, for example, is an important example of such a mathematical structure in logic.

Similarly to the personal connection that took place between André Weil and Claude Lévi-Strauss in New York, an encounter took place between Jean Piaget and Jean Dieudonné, and this meeting also affected the development of science in a profound way and exerted Bourbaki's influence on the progress of structuralism. The meeting took place in April 1952 at a conference on Mathematical Structures and Mental Structures, held outside Paris. Dieudonné gave a talk in which he described three of Bourbaki's "mother structures": composition, neighborhood, and order. Then Piaget gave his talk, in which he described the structures he had found in children's thinking. To the great surprise of both speakers, it was clear that the two of them spoke about exactly the same ideas. It was evident that there was a direct relationship between the three mathematical structures studied and promoted by Bourbaki and the three structures inherent in children's operational thinking.[2]

At the Cerisy-la-Salle conference in 1959, mathematics was not emphasized at the beginning of the talks. But on the second day, Jean Piaget brought it up very forcefully. The structures pioneered by Bourbaki were the key elements stressed by Piaget, who summed it all up by saying that systems presented laws that were totally distinct from the laws governing the behaviors of single elements.[3] Piaget explained elsewhere that the roots of the spontaneous psychological development of arithmetic and geometric operations in the minds of children paralleled the concepts used by mathematicians. He referred to the "linear order" of science as extolled by the positivist thinkers of the Vienna Circle, arguing that it needed to be replaced by a mathematical circle.[4]

For Piaget, the mathematical sciences were the basis for all science—including the "science of man"—and, in turn, mathematics itself was based on hidden structures of the human mind.[5] Bourbaki's structures were the key elements Piaget was looking for in understanding the inner workings of the human mind. These structures determine both the workings of the brain, and, through the effects of the brain on social behavior, they also determine the behavior of an entire society—as shown by Lévi-Strauss's work.

Piaget's "genetic structuralism" appeared as a unified methodology that was equally adaptable for use in a wide variety of fields other than psychology. He saw in mathematical structuralism the answer to questions posed by all science. In mathematics, he commended Bourbaki for bringing structuralism to the forefront of all mathematical analysis, making it widely applicable in all forms of human investigation. He saw Bourbaki's structures as deeply rooted elements of the human brain.[6]

A meeting was held in Paris from April 18 to 27, 1956, the theme of which was "Notions of Structure and Structures of Knowledge." The idea here was to synthesize knowledge across disciplines with the concept of structure. The organizers of this meeting hoped to exhibit "an isomorphism between different sectors of knowledge," thus borrowing from mathematics the key concept of isomorphism: a structure-preserving map. An isomorphism is a function from one set into another that also preserves the *relationships* among the elements of the set that is being transformed. This way, when we look at the set that is the result of the *mapping* by the isomorphism of the original set, this new set preserves all the relationships that are inherent between objects of the original set. But the conference had overly ambitious goals, and ultimately these were not met. More questions were raised than answered and the conclusion was that "no solution has been found; the structure of knowledge has not been defined."[7]

The question remained of whether Bourbaki's definition of structure really meant anything outside of mathematics. One form of structuralism that developed during those years was divorced from mathematical thinking altogether.

WHEN LÉVI-STRAUSS MET up with Jakobson in New York and began to study structural ideas in the context of anthropology, Jakobson was very interested in that work. On his own, he had deduced that if the structural elements are embedded in all language, and hence in the human mind, then social behavior as studied by the anthropologist should also reveal elements of deep structure. Lévi-Strauss, too, understood this idea and pursued it with much vigor. It was because of his great success in uncovering the elements of structure within systems of kinship, thus exposing their importance beyond the field of linguistics, that Lévi-Strauss is widely acknowledged as the initiator of structuralism. He had studied Freud since his teens, and was always interested in psychology and psychoanalysis. Once he developed the ideas of Jakobson and took them out of the confines of linguistic analysis, Lévi-Strauss sought to apply these structuralist methods in the field of psychology, in addition to anthropology.

The genius of Claude Lévi-Strauss was that he was able to seize the rudimentary ideas proposed in linguistics by Roman

Jakobson, based on the origins suggested by Troubetzkoy, and to realize that they were mathematical principles that could be applied with great generality. His encounter with the ideas of Bourbaki on structuralism and the structural roots of mathematics through his encounter and joint work with André Weil further spurred him on.

Lévi-Strauss hypothesized that the mathematical kernel of analysis—the structural root of all systems—was embedded in the human mind. He came to believe that the human unconscious was attuned to these basic structural elements, and that what scientists find in the analysis of the brain through psychoanalysis, linguistic analysis, and anthropology are actually the basic elements of human thought. Mathematics, therefore, is deeply embedded in the human brain, something that would be rediscovered toward the end of the twentieth century through the work of psychologists studying how the brain does mathematics.

BUT THE PERSON who made the most important use of structural ideas in psychiatry and psychoanalysis was French psychiatrist Jacques Lacan (1901–1981). Lacan exploited the connection between structuralism and the unconscious mind. The idea of a latent structure inside the brain is reflected today in methods in the area of artificial intelligence: neural networks and genetic algorithms. Both of these methods of applied mathematics used in artificial intelligence assume that there is a latent level to thought processes. What artificial intelligence tries to do is to mimic in a computer model what the brain is believed to be doing. The use of such methods attempts, therefore, to uncover the hidden level of reasoning that operates within the human brain and to use it. But the idea that there is such a hidden level in human thought processes goes back to structuralism. Jacques Lacan tried to apply the structuralist approach to identifying the hidden level in the human brain. To him, that hidden level, the latent structure within human thought, was the subconscious.

Lacan started his life by rejecting the circumstances and values of his upbringing. The child of a deeply religious Catholic family, he abandoned religion and embraced philosophy, with special interest in the work of Hegel. He studied neuropsychiatry and began to practice it, and then he switched to psychoanalysis.

Lacan broke away from the French association of psychiatrists by rejecting the norms accepted by the organization for the practice of psychoanalysis. For example, he refused to conduct sessions of prescribed length with his patients, instead choosing to determine on his own the length of time each therapy session should last. He studied the ideas of Freud and reread them with a structuralist interpretation. Freud's view of the subconscious mind was reinterpreted by Lacan as the seat of the hidden structure of the brain, in which symbolism reigns. Lacan was directed by his research of these structures to surrealism and myth. He made the acquaintance of Dali and studied the artist's surrealist paintings in search of the structures of the mind.

Lacan studied paranoia and psychosis and attempted to identify the structural elements within these illnesses using a Freudian approach. He used novel ways of exploring these ideas of structure in psychoanalysis, for example, by "the mirror experiment," in which very young children were shown their own reflections in a mirror for the first time and their responses recorded. Lacan analyzed his data using the assumption of hidden structure. This led him to results that confirmed the structuralist approach.

Freud had described the self as divided between the conscious and the unconscious. The unconscious drives the desires, while the conscious self's actions, thoughts, and beliefs reside within the conscious part of the mind. Lacan began to think about these Freudian concepts and developed his own ideas, which can be seen as based on Freud's notions together with the structuralist ideas of the linguists and Lévi-Strauss. Freud's thinking already had some structuralist germs because he questioned the humanist ideas of free will and self determination, believing that the subconscious—the hidden level in the mind—contributed to behaviors. But Freud believed that by bringing the subconscious into consciousness, he could minimize repression and, thus, neuroses. The id can be replaced by the "I," Freud hoped, thus eliminating the "structuralist" hidden level of human thought; the effects of the unconscious component can be minimized. But according to the theory that Lacan developed, the ego can never suppress the unconscious or control it, because the unconscious—the structuralist center of the mind—is the central element of the brain. The ego is thus seen as an illusion, a product of the unconscious mind.

In Lacan's Mirror Stage experiment, he described the process by which, he believed, the infant forms an *illusion* of an ego or self, a unified consciousness that arises when seeing one's own reflection. In reality, the infant only has an unconscious, and the conscious is an idea formed within the ever-present unconscious. The central concept of the unconscious mind, according to Lacan, is structured like a language. Here is where Lacan uses the structuralist ideas developed by linguists. Freud postulated two main mechanisms that governed unconscious processes: condensation and displacement. Meaning is thus either condensed into metaphor, or it is displaced in metonymy.

Lacan saw both of these processes as essentially linguistic in their nature. He viewed Freud's analysis of dreams as well as the analysis of unconscious symbolism used by patients as verbal in nature. Therefore, the unconscious mind, according to Lacan, is keenly attuned to linguistic structure. The elements of the unconscious—wishes, desires, images—all form signifiers in the mind, according to Lacan's theory. These signifiers are linked in a chain, and there are no signified entities. The chain of signifiers in the brain keeps shifting. One signifier always leads to another signifier, in the same way that in a dictionary one word leads to another. The unconscious mind, according to this theory, consists of a continually circulating chain of signifiers with no anchor.

Lacan thus used linguistic ideas to recast Freud's picture of the unconscious mind into an ever-shifting chain of signifiers: drives and desires that lead from one to another. According to Freud, bringing the chaotic drives and desires into the conscious mind is the process by which a child becomes an adult. According to Lacan's theory, however, this is impossible. Bringing the unconscious signifiers to the level of consciousness is simply an illusion. The child becomes an adult by entering a symbolic stage through language.

Lacan's unorthodox ideas were derived through the influence of his reading of Saussure, the pioneer linguist of the turn of the century. But these ideas evolved more in keeping with the structuralist approach after Lacan read Lévi-Strauss and, through his ideas of structure in anthropology, came to Lévi-Strauss's linguistic origins in structuralism.

In his major writings in the mid-1950s, Lacan quotes

generously from the linguistic literature. In Paris, he was often visited by Jakobson, whose ideas he incorporated into his work. "It is all the structure of language, which the psychoanalytic experience uncovers within the unconscious," he wrote.[8] Carrying the idea of symbolism in the brain to a higher level of structural abstraction, Lacan adopted a technique used by Saussure and other linguists, which transcribes the relationship between signifier and signified in a mathematical formula. The Freudian condensation, for example, created a functional metaphor. Here, the ratio of the signifier to the signified is written as a mathematical fraction, with the signifier in capitals and the signified in lower-case. It is the image "divided by" the actual concept:

Lacan thus took the structural ideas brought over from linguistics and applied them rigorously, creating precise scientific concepts in psychology and psychoanalysis. This was the next stage of abstraction within a discipline that until structuralism was less precise and used vaguer concepts.

Lacan followed the ideas and work of Bourbaki. He not only liked what Bourbaki was doing in terms of promoting structural thinking in mathematics and exploring new mathematical ideas, but he also liked the *way* Bourbaki worked. At some point, therefore, Lacan championed the establishment of a group of psychiatrists and psychoanalysts that would be based on the model of Bourbaki. This would be a secret society that would publish works in its discipline under a common pseudonym. Unfortunately, his idea did not take root.

THE AREA OF economics has always enjoyed greater precision and wider use of mathematical concepts than do social sciences such as anthropology and sociology. And yet, the advent of structuralism, and in particular the ideas of Claude Lévi-Strauss, brought new thinking into the discipline. In particular, economists in the 1950s were interested in applying structural ideas to their science by better incorporating the concept of a *Model* in the spirit of its use in anthropology.

The concept of economic equilibrium is an excellent example of axiomatization in economics, and of the construction of a structuralist model that works exceptionally well. Since structures discovered in linguistics, anthropology, and psychology are manifestations of human behavior that is completely dependent on latent structures inside the human brain, it stands to reason that economic behavior—which is the aggregate behavior of large groups of human beings—should exhibit similar structure.

Models of economic equilibrium capture this idea, and their great success can be seen as a major proof of structural thinking in all the sciences that deal with human behavior. A model of economic equilibrium postulates the existence of two curves: a supply curve and a demand curve. Such mathematical functions do not "exist" in the real world—they are simply models of human behavior. The first curve says that as the price goes up, the supply will go up. And the second says that as the price goes up the demand will go down. The equilibrium point, which determines the actual price of a commodity in a given free economy, is the intersection of the two curves. The price is determined so that supply and demand meet.

The supply-demand equilibrium model is one of the best examples of structuralist thinking. It shows how mathematical ideas can be used to model the workings of the human brain, viewed through the aggregate behavior of large groups of people. The fact that the model works so well is a good

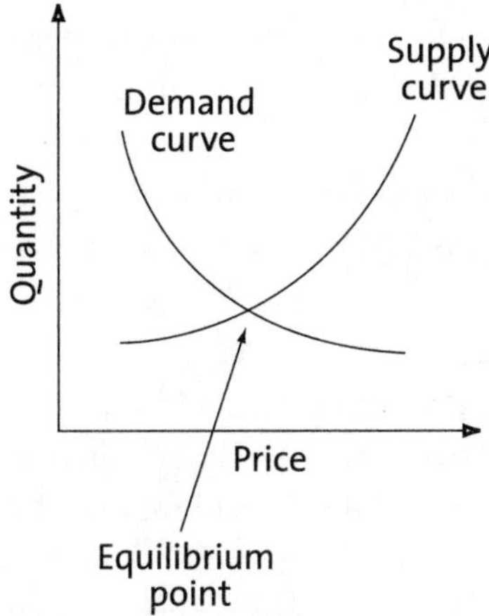

indication that it does indeed capture the symbolic, latent functioning of brain processes. Economics, in this sense, is also a "language," and its ideas and symbolisms are as amenable to manipulation using symbolic logic as are the concepts of linguistics. There is a similar analogy between the workings of an economy and the structures of kinship in a society, the code for which was broken by Lévi-Strauss.

Economic modeling goes far beyond the supply and demand equilibrium model. Already in the late 1940s, the American economist Wassily Leontief (1906–1999) attempted to construct a model that would describe the workings of the entire American economy. Leontief's model was based on the economic activity of the largest American corporations. His assumption was that these corporations interact with each other and with the general public, and that a mathematical model should be possible for capturing the nature of these interactions, and thus predict the behavior of the entire economy. The model enjoyed success, as did further models that Leontief developed, and he was later awarded the Nobel Prize.

Once econometric modeling, as these mathematical tools are called, entered the mainstream of economic research, wider model-building activity ensued. Today, much of what we learn about the economy comes from such models, and it is important to understand that the underlying philosophy that drives the construction and use of such models is structuralist thinking.

Since Bourbaki's structures are mathematical, there is a common link between them and the structural models used in economics. And while model building developed on its own in the field of economics, rather than through the instigation of Bourbaki, the relations between mathematics and economics have continued to reinforce ideas in both fields. Often, important ideas in mathematics find their first applicability outside mathematics (and physics) within the area of economics. Thus, for example, topology, abstract algebra, and the theory of equations have all found important uses in economics. But Bourbaki's influence reached far beyond the social sciences, linguistics, and economics. It extended to philosophy and literature as well. In literature, the foundation and work of the group called Oulipo is a prime example of the direct influence of Bourbaki.

thirteen

THE LITERARY GROUP OULIPO

OULIPO IS A literary society loosely modeled after Bourbaki. The Oulipo membership—including a few well-known European and American writers—adopt a number of Bourbaki's goals and principles. One of the group's goals is to reshape literature in a new way; another is to break away from accepted norms of writing; and a third is to follow a seemingly lighthearted approach to their oeuvre.

On November 24, 1960, the semisecret society of Oulipo was founded by François Le Lionnais and Raymond Queneau, who was a member of the French Mathematical Society. The Oulipo society sought to experiment with formal constraints imposed on the production of literature. In an interview on French radio in 1962, Queneau defined "potential literature" as follows:[1]

> The word potential concerns the very nature of literature. That is, it is less a question of literature, strictly speaking, than of supplying forms for the good use one can make of literature. We call potential literature the search for new forms and *structures*—to use this

> slightly learned word—that may be used by writers in any way they see fit.

In his "Second Manifesto" of the society, Le Lionnais emphasized that the structures the Oulipo group concentrated on were those of mathematics, as promoted by Bourbaki. Oulipo was interested in methods of literature rather than in specific examples or cases. This is reminiscent of Bourbaki's distaste for applications.[2] The relationship between this literary group and Bourbaki was very strong.

In addition to Queneau and Le Lionnais, two other members of Oulipo, Claude Berge and Jacques Roubaud, were also mathematicians. And Queneau, in fact, visited a Bourbaki congress, the one held in March 1962. Oulipo and Bourbaki both emphasized axiomatics, formal beauty, and–humor.[3] Oulipo is the French acronym of Ouvroir de Littérature Potentielle (Workshop for Potential Literature), which began with the work of Raymond Queneau (1903–1976) and François Le Lionnais (1901–1984). The two had the idea of setting a foundation for literature analogous to Bourbaki's geometry. To transfer these ideas to literature, they would make the following transformations of geometrical ideas to linguistic ones:

Point→Word
Line→Sentence
Plane→Paragraph

The Oulipo group thus organized themselves with the purpose of promoting these ideas and developing literature using analogs of the revolutionizing structural principles of Bourbaki.[4] Much of Oulipo's oeuvre has to do with mathematical and quasi-mathematical ideas of combinations, permutations, and Fibonacci sequences as well as random structures. Oulipo attempted to deconstruct literature and rebuild it in a new way, thus bringing the new structuralism from mathematics, anthropology, and art into the foundation of a new literature.

Like other groups of the twentieth century—the surrealists and Bourbaki—the Oulipo group wrote a manifesto. In fact, Le Lionnais, the founder of the group, put forward *two* manifestos. In the first manifesto, titled *La Lipo,* Le Lionnais wrote:

> Mathematics—and particularly abstract structures of contemporary mathematics—offers us a thousand directions of exploration, starting with algebra (use of new laws of composition) and topology (considerations of neighborhood, openness or closeness of texts). We also dream of anaglyphic poems, of texts transformable by projection, etc. Other directions are imaginable as well, notably in the realm of particular vocabularies (crows, foxes, porpoises; the Algol language of computers, etc.). It would take a long article to enumerate the possibilities that can be made out from here, and some that can only be sketched out."[5]

The founders of the group clearly realized the potential usefulness of mathematics, especially as newly axiomatized by Bourbaki, for creating novel ways to produce literature. In true Bourbaki fashion, Oulipo began its career on Thursday, November 24, 1960, in the cellar of a restaurant named Le Vrai Gascon. Here met François Le Lionnais, Raymond Queneau, Jean Queval, Jean Lescure, Jacques Duchateau, Claude Berge, and Jacques Bens. In the transcript of the meeting, the seven founders wrote the following: "We are meeting here not only to entertain ourselves (that which is already considerable, certainly), but also to answer the question: what results may we expect of our works?"[6]

Again as in the case of Bourbaki, a number of other members joined the group, among them some non-French members. The most prominent of these latter was the Italian writer Italo Calvino. At first the group named itself the Seminar for Experimental Literature (in French, Séminaire de literature expérimentale) and took the French acronym SLE

A month later, at the group's second meeting, on December 19, 1960, the name was changed to Oulipo. A new member who had just joined the group, Albert-Marie Schmidt, suggested the name change. First the acronym was Olipo, but the members decided that Oulipo sounded better. (In French, "ou" is pronounced as "oo.") The members of the group liked their new name so much that they wrote many poems based on it. Here is one of them:[7]

Hou! Lippe, eau!
Où Lipp? Haut?
Houx lit: "peau"
Houle hippo!
Où lit, pot?

[Translation:
Boo! Lower lip, water!
Where Lipp? High?
Holly bed: "skin"
Swell hippo!
Where bed, pot?]

The Oulipo writers composed thousands of poems and sonnets, many of them based on mathematical concepts of combinations and permutations of nouns, verbs, and letters. One of their most successful methods of creating poetry was what they called the "S + 7" method, which they considered a special case of the "M +/- n Method." This was a system of substitution by which a known piece of poetry, or any piece of text, was changed as follows:

Choose a text. This could be anything; in their first example, Oulipo used a May 6, 1961, article from the French newspaper *Le Monde*.

Choose any dictionary you like. Oulipo liked to use French-English and French-Italian dictionaries for this purpose.

In the S + 7 variation of the method, replace every noun with the seventh noun appearing after it in your dictionary (in the original language of the text). In the M +/- n generality, choose the nth noun (e.g., if n=7, choose the seventh noun) after each noun in the text, or the nth noun appearing before each noun in the text. In the general method, the same can be done with verbs, adverbs, adjectives, and so on.

Oulipo developed sophisticated variations of this game, using various French texts, including statements of Euclid's theorems in French. Of course, good knowledge of French is necessary for understanding the resulting "experimental literature" that resulted from these methods. But the reader is invited to play these games using any text in English, or in any language desired. The results can be quite amusing. And they teach us something about the structure of language.

Raymond Queneau introduced ten sonnets of fourteen verses each. Queneau gave rules by which the reader could exchange each verse of a sonnet by one of the nine other variations corresponding to the verse. Thus each verse has ten possibilities, and since there are fourteen verses in each sonnet, the total number of possible sonnets is ten to the fourteenth power, which is one hundred thousand billion.

This, in fact, was the name Queneau gave his work, in French: *Cent mille milliards de poèmes.* He noted that his mathematical way of producing poetry was exponential. The number of possible poems grows exponentially with the number of verses. Queneau then computed the time it would take a person to read all his sonnets. Making some assumptions about the speed at which one could read the sonnets, reading eight hours a day, two hundred days a year, he came up with the answer: 190,258,751 years.[8]

Oulipo also ventured into Kabbalah, where they found the mystical ancient art of forming and numerically evaluating combinations of letters and words. Another one of their areas of interest was the construction of poetry and other texts with or without certain letters. Here, they used foreign writings as well, in particular some poems written in English. They presented two nineteenth century anonymous poems in English. The first one does not contain a single letter *e,* but every other letter of the alphabet appears in the poem at least once:

Quixotic boys who look for joys
Quixotic hazards run
A lass annoying with trivial toys
Opposing man for fun

In the second poem, each line uses only a single vowel:

War harms all ranks, all arts, all crafts appal!
Idling, I sit in this mild twilight dim

Whilst birds, in wild swift vigils, circling skim
Bold Ostrogoths show no horror of ghosts
Lucullus snuffs up musk, Mundungus shuns

They also pointed out that an American sailor, Ernest Vincent Wright (1872–1939), published a book the year he died titled *Gadsby: A Story of Over 50,000 Words Without Using the Letter E.*[9] Oulipo produced many poems in French following this model—not containing the letter *e*—and experimented with the avoidance of other vowels as well.

Another kind of substitution method studied by Oulipo consisted of exchanging the words within each sentence. This had the effect of reversing or changing cause and effect and revealing hidden meanings that a piece of poetry could have. One example of this idea is the following interchange of words:

Calm rain causes idleness
Idleness causes calm rain

Possible exchanges of words in this sentence are as follows.[10]

rain calm causes idleness
Rain causes idle calmness
Idle calmness causes rain
Calmness causes idle rain

Oulipo strove to use other mathematical methods in literature, and to apply terms and ideas of mathematics; in particular, they attempted to use the following mathematical concepts:

> Set, class, element, inclusion, exclusion, union, intersection, complementation, symmetric difference, associativity, commutativity, neighborhood.

Thus the ideas of Bourbaki and its structuralist approach in mathematics—basing everything on set theory and elementary concepts—found its way into literature through the work of Oulipo.

In topology, the group adapted the Möbius strip for its use. Members wrote poems on the surface of a Möbius strip and experimented with the outcome. Computers were just beginning to be available at the time the group began its work in the early 1960s, and computer languages, especially ALGOL, were receiving much attention from the public. There was general wonder and amazement at the way computers were made to act using "commands" that resembled human languages. People were fascinated by these ideas and wanted to understand the relationship between human language and computer languages. Oulipo was especially attracted to ALGOL and its structure and used this computer language—and often the attendant Boolean logic of computers—to create new poetry. Here is one such poem (translated from French by the author):

> Start: To make etiquette
> Go To commentary
> While Variable Not False
> Otherwise True. End.[11]

Oulipo was well aware of the ideas it adopted not only from Bourbaki but also from Lévi-Strauss's pioneering structuralist work in anthropology, and they mentioned this work in their writings. Their stated purpose was to apply Lévi-Strauss's very successful structuralist methodology, combined with Bourbaki's mathematical work, to create a new literature. As François Le Lionnais expressed it in a meeting of Oulipo, held in his garden on a warm August day in 1961, the purpose of Oulipo was to provide future writers with new techniques that could help them explore ways of producing literature. These new methods, pioneered by Oulipo, should give writers a greater freedom to explore the immense possibilities of their art.

Oulipo held very frequent meetings over the decade and a half that followed. They explored new ways of producing literature using a wide array of techniques borrowed from mathematics and other fields, and they wrote a treatise that described and exemplified these new ways of producing literature. By the late 1970s, however, the group began to decline. But it did leave us an interesting legacy and a body of work that resembles that of Bourbaki.

fourteen

ALEXANDRE GROTHENDIECK AND THE IHES

AFTER A FEW years in Paris, Alexandre Grothendieck had to leave the French capital, a city he had learned to love and in which he had hoped to settle permanently. Having no citizenship and being able to travel only on a UN passport, Grothendieck found it very hard to obtain a position in France that would provide him a livelihood. Doing mathematics was not enough if one could not earn a living doing it. He therefore moved to Brazil, and taught courses on topological vector spaces in São Paulo during 1953 and 1954. He then spent the first part of 1955 at the University of Kansas, in Lawrence, Kansas. He went on to visit the University of Chicago for the rest of that year. All this time, Grothendieck was hoping to find a position in France but continued to experience great difficulties doing so because of his citizenship problems.

While visiting Rutgers University in New Jersey during his American tour, Grothendieck met a young graduate student in mathematics named Justine. She left her program in order to travel with him, and moved with him to France when he returned to that country to try to obtain an academic position once again. They lived together in France for two years and had a child. That son is now a mathematician, having recently received his PhD in statistics.

ACCORDING TO MATHEMATICIANS who knew him, Grothendieck is modest and perhaps even "naïve and childlike," and certainly does not display any arrogance. He has a great capacity to enjoy life, is charming, and loves to laugh.[1] He always enjoyed the simple things in life, and has been described as wearing sandals made of old tires. He has also been known to shun rugs in his home, considering them a decorative luxury for frivolous wealthy people.[2] At the end of 1957, his mother died. Grothendieck, who had been exceptionally close to his mother and greatly admired both her and his father, underwent a shock, which caused him to leave mathematics for a few months. He returned to mathematical research in 1958, and soon afterwards married a woman named Mireille, who had been a close friend of his mother and was several years older than he. Eventually he would have three children with her.

In 1957, Pierre Cartier made an amazing observation. He understood that a ringed space locally isomorphic to a ringed space of the form Spec (A)—the set of all prime ideals of a commutative ring A—should be considered as a generalization of an algebraic variety. Cartier told Grothendieck about this idea, and the latter began to develop the foundations of algebraic geometry based on Cartier's generalized notion of algebraic variety, now known as a *scheme*.

Grothendieck then began to plan a very ambitious project: to rewrite all of algebraic geometry in a book titled *Eléments de Géometrie Algébrique*. He saw the book as comprising thirteen chapters, in which much of algebraic geometry would be developed in the new language of schemes. Grothendieck was prescient enough to realize that his new approach could be instrumental in proving the conjectures made some time earlier by André Weil about the zeta function of algebraic varieties over finite fields.

Grothendieck planned to devote the final chapter of his book to this proof. But he was able to finish only four chapters of the ambitious project; those four numbered around 2,000 pages. This work was carried out jointly with Jean Dieudonné. A seminar on these topics, which Grothendieck gave during the 1960s and early 1970s, provided the details that would have been included in the remaining, unwritten chapters.[3] Grothendieck's book and seminars exerted an immense impact on the development of mathematics in the second half of the twentieth century. His ideas paved the way for one of his students, Pierre Deligne, to eventually prove the Weil conjectures. Furthermore, Grothendieck's innovations in the field were instrumental in Gerd Falting's proof of the Mordell conjecture—the difficult problem Hadamard had wanted Weil to solve as his dissertation topic. These same ideas were of paramount importance in the eventual proof of Fermat's Last Theorem by Andrew Wiles in the early 1990s.

One of Grothendieck's great early achievements in his project was making commutative algebra a part of algebraic geometry. Grothendieck thus established the role of commutative algebra as the study of the local structure of schemes. The theory of schemes also weds arithmetic with geometry, thus achieving a goal set a century earlier by German mathematician Leopold Kronecker. But the theory of schemes not only added new objects of study to algebraic geometry, it also brought new insights and instrumental techniques to the quest of solving important problems in classical geometry.[4]

Grothendieck's monumental work brought us new insights

into the study of the very nature of space and its points. Euclid saw points as the basic elements of geometry. Points produce lines and circles; and the intersections of lines, or lines with circles, give us points. But a point was something that Euclid defined axiomatically: a point was something which had no length or width or breadth. It was something that was defined by what it was not.

In the seventeenth century, Newton and Leibniz took up the reexamination of space and its points. For Leibniz, the constituent of all things—spiritual as well as physical—was something he called a *monad*. The monads, reflecting Euclid's points, were "windowless" elements, meaning that they had no internal structure whatsoever, and the only interesting property they possessed was the relationships they had with one another.[5]

On the other hand, the definition of points and sets that Bourbaki chose to use in its first published volume of the *Éléments de mathématique*, the volume on set theory, is as follows:

> A *set* is composed of *elements* capable of having certain *properties* and having certain *relations* among themselves and with elements of other sets.[6]

Thus, points are seen as preexisting, and the problem of mathematics is to organize them and give them structure. Grothendieck, inspired by the ideas of the German mathematician Bernhard Riemann (1826–1866), who discovered the notion of a surface stacked over a plane, proposed the abstract idea of *topos*. Grothendieck came to the idea of topos by replacing the open sets of a space (the basic elements in topology) by spaces stacked over the given space. The same idea can be viewed by considering the *category* of sheaves over the space. These notions, in fact, came from category theory, which was discovered and developed in the 1940s by the American mathematician Saunders MacLane and by Bourbaki member Samuel Eilenberg.

The main idea of category theory is to consider general objects and their transformations, rather than points. Category theory is a very abstract discipline, in which the nature of the objects one deals with does not matter; it is only the relationships among these objects that are of importance. For example, the painting below, by the artist Thomas Barron, represents houses. But the nature of the objects—the houses, in this case—has no significance. The artist is only concerned with the mutual relationships among these objects—in this case, how these objects relate to each other spatially. This same idea is the key concept in category theory, where objects have no meaning on their own, except in the way they relate to one another.

Thomas Barron, *Promontory—Swampscott, III.* Gouache and charcoal, 1986 (REPRODUCED BY PERMISSION)

The concept of *topos*, the abstract idea derived from notions in category theory, was for Grothendieck the ultimate generalization of space. Grothendieck then claimed the right to transcribe mathematics into any topos he might choose.[7] Using these new ideas, Grothendieck was able to bring parts of modern mathematics to great heights that Euclid, Leibniz, Riemann, or even his own contemporaries could not even dream of. The new level of abstraction and the new and very general way of defining space allowed Grothendieck to prove important results in mathematics, ones that were out of reach for anyone before these amazing developments took place. What Grothendieck did was to inject into geometry the powerful methods and concepts of abstract algebra.

These concepts, now viewed in the context of space and geometry, made many new things possible in mathematics. They redefined space itself, and they allowed a mathematician to see things in an entirely new way. They also united various branches of mathematics, bringing a new understanding of mathematical ideas that would not have been possible otherwise.

It was this kind of amazing insight into problems that people have looked at before but could not solve that made Grothendieck's achievements so astounding. And it is because of his great insight and understanding that he can be compared with Albert Einstein.

The theory of categories provided a suitable framework for describing general properties of objects studied by mathematicians—a framework that, following unsuccessful attempts, Bourbaki decided not to include within its own work.[8] Category theory was like a superstructure hanging above set theory, abstract algebra, topology, and other areas.

This superstructure contains the essential elements of a mathematical theory, which can then be applied to sets, algebra, topology, and so on. It is thus a very powerful theory. But Bourbaki had made its decision back in the 1930s to base all its work on set theory. It did not now want to go back and base its oeuvre on category theory.

LATE IN 1957, a French businessman with a strong interest in physics decided to fund the establishment of an institute near Paris to be modeled after the Institute for Advanced Study at Princeton. This would be a top-notch center for research in mathematics and theoretical physics. It was named the Institut des Hautes Études Scientifiques (IHES)—the institute for high scientific study. The institute opened in June 1958, and sometime later it moved to a hill in the countryside southwest of Paris.

Today, the IHES is one of the world's most prestigious institutes for mathematics research. It has been argued that the institute was actually created so that Grothendieck, who had no nationality and no prospect of finding a position in France, could have a place of employment. As soon as the institute was founded, Jean Dieudonné and Alexandre Grothendieck were jointly appointed to the two professorships in the institute. They took up their positions in March 1959, and Grothendieck started giving seminars on algebraic geometry. This was then a new field into which Grothendieck moved, having earlier made his stunning contributions to topological vector spaces and functional analysis. But his greatest work was still ahead of him—to be done here at the IHES. Grothendieck completely reworked the field of algebraic geometry. He also wrote new books on mathematics, treatises that would compete with the books written by Bourbaki.

Grothendieck's road in mathematics began to diverge from that of Bourbaki. Finally, Grothendieck and Bourbaki had a falling out. Some mathematicians have surmised that this rift was due to an antipathy between Grothendieck and André Weil.[9] This is likely, since Weil was a somewhat jealous person who clearly saw that Grothendieck was a far better mathematician than he was. Nor did Weil want to lose his control over the group, and Grothendieck was pushing the membership in new directions in mathematics.

Grothendieck did not much like the system of Bourbaki, in which every detail had to be discussed and argued about. His coverage of topics in mathematics was deeper, and it focused on particular areas, rather than following the very general approach of Bourbaki, who wanted to cover every area in an encyclopedic way. In 1960, the final break took place, and Grothendieck—frustrated and angry with many of the members—left the Bourbaki group for good. But he remained in contact with some of the members. These people were, after all, among France's greatest mathematicians, and he wanted to continue to interact with them.

Alexandre Grothendieck was the most gifted French mathematician of the 1960s. Grothendieck used his deep insights to discover hidden relationships between mathematical objects. These relationships then revealed to him hidden aspects of the objects he studied. Grothendieck invented—or discovered—new mathematical constructions. One of the most important of these was the deep concept he termed a *motive.* The word, in French, actually means pattern, rather than motive, and Grothendieck's motive is a hidden pattern within a mathematical structure in the field of algebraic geometry. Grothendieck described in his writings his aim of unifying two worlds: "the world of *arithmetic* in which live the spaces with no notion of continuity, and the world of *continuous size*, in which live the spaces in the proper sense of the word, accessible to the methods of the analyst."[10]

The ideas in his work have to do with the mathematical objects called *schemes, sheaves,* and *topos.* Sheaves were conceived by Jean Leray and later developed further by Henri Cartan and Jean-Pierre Serre. The new theory, based on Grothendieck's topos holds perhaps a great promise of replacing the traditional theory of sets, with its inherent paradoxes.

A scheme is a generalization of the concept of variety. One remarkable aspect of Grothendieck's work was his introduction into mathematics of ideas of great generality. His generalization of a variety into a scheme allowed people to understand all the incarnations of a variety in many different settings.

In this sense, Grothendieck is the greatest structuralist the Bourbaki group ever could count among its members. He is a giant in mathematics who thinks in great generalities, has incredible vision, and can foresee the right course to take in research. Grothendieck always knew which paths of research in mathematics held the greatest promise of important results. His problem in life was to convince other people that his direction was the right one.

GROTHENDIECK TRAVELED FOR a few years, and held visiting positions at Harvard and other American universities. When he returned to Paris in 1961, he resumed his work and his seminars at the IHES. His seminars attracted many mathematicians from France and around the world. This became the "Grothendieck school," in which his ideas were developed, promoted, disseminated, and studied. Grothendieck had a wonderful ability to match people with problems. He somehow could sense which particular problem would best be assigned to any given student or colleague, and many of his ideas then became theorems and dissertations for his students.

For a dozen years, Grothendieck was the undisputed master of algebraic geometry. His vision for this discipline was more than could be accomplished by one mathematician alone. So he parceled out his insightful ideas to many around him, who then vigorously pursued solutions to the problems he had proposed to them. A number of leading American, Russian, and Japanese mathematicians had had their research take off because Grothendieck encouraged them to pursue the right problems and gave them his best ideas and good guidance.

Grothendieck introduced into the areas he studied great precision in the analyses he carried out and the concepts he derived. He championed a trend toward increasing generality and abstraction in mathematics.[11] He did not pursue generality in a frivolous way, but rather because it allowed him to see things clearly. His mind was so different from those of other people that concrete examples simply did not work for him. Usually, in mathematics one works with examples and then tries to generalize them.

Grothendieck is very different here: His mind works on generalities. There is a story that once, at his seminar, Grothendieck said something about a prime number, and one of the participants asked him if he would make it more specific. "You mean give an actual prime number?" Grothendieck asked, and the student replied, "Yes, please." So Grothendieck said, "Well, then, take fifty-seven." Of course, fifty-seven is not a prime number (it is nineteen times three). Was Grothendieck joking, then? Surely such a great mathematician would have known that fifty-seven is not prime. But according to David Mumford of Brown University (quoted in Jackson, 2004, p. 1197), Grothendieck did not necessarily know that. "He simply doesn't think in concrete terms." Grothendieck is interested in the general properties of numbers, not in any particular number. The number fifty-seven is now affectionately called Grothendieck's prime.

This quality of Grothendieck's has been contrasted with those of the Indian mathematician Srinivasa Ramanujan (1887–1920), who was believed to have had a personal acquaintance with every number. There is a well-known story about Ramanujan. When he was lying ill in a hospital in Britain, an illness from which he would not recover, he was visited by his friend and benefactor, the famous English mathematician G. H. Hardy (1877–1947). Hardy mentioned to Ramanujan that he had arrived by taxi, and the taxi had the very boring number 1729. "No, Hardy! No, Hardy!" said Ramanujan. "It is a very interesting number. It is the smallest number expressible as the sum of two cubes in two different ways!" ($1729=10^3 + 9^3 = 12^3 + 1^3$). This is not how Grothendieck thinks: his mind operates in sweeping generalities and great abstraction.

Grothendieck also appeared in the regular "Bourbaki Seminar," which continues to this day. But once Grothendieck left Bourbaki, both the seminar and the group were weakened. Bourbaki's greatest error was letting Grothendieck go and disagreeing with his vision for the future of mathematics.

In 1966, Grothendieck was awarded the Fields Medal for his great contributions to mathematics. Since mathematics has no Nobel Prize, the Fields Medal is generally considered its equivalent in mathematics. This is the highest honor mathematicians can bestow on their own. Grothendieck was to be awarded the Fields Medal at the International Congress of Mathematicians, to be held that year in Moscow. For political reasons, however, he refused to travel to Moscow to receive the award. He had become very political in his views and activities, and these political preoccupations were also taking time away from his mathematics.

Grothendieck was getting burned out from working too extensively on mathematics day after day, night after night. Then, in 1968, the Soviets invaded Prague, and in Paris students carried out massive street protests. As a result, millions in France went on strike and the French government, fearing a revolution, stationed tanks around the capital. Grothendieck became increasingly political. Some have said that the riots of 1968 ruined mathematics forever for Grothendieck, and that following the events of 1968 he began to devote virtually all his efforts to political action in the environmental and antiwar movements.

He traveled to Vietnam to protest the war; he participated in many peace rallies; and he refused to accept any government research grant, fearing that by doing so he would tacitly be supporting political causes he abhored. He would talk about politics whenever invited to give a lecture on mathematics. This practice irritated all who had come to mathematics meetings to hear him speak—regardless of whether or not they agreed with his politics. His listeners did not want to hear political tirades, delivered in a very aggressive way, rather than mathematics.

Pierre Cartier, who knew him well, described Grothendieck's behavior during this period:[12]

> The student uprising of 1968 (in France) was to a large extent the result of [the anti-Vietnam War] movement. Its anarchist aspect held attractions for him and forced him to admit that he had ceased being an outcast and had become a scientific mandarin. The movement of May 1968 aroused others among the Bourbaki group, such as Chevalley, Samuel, and Godement. The cold war was at its height, and the risk of a nuclear confrontation was very real. The problems of overpopulation, pollution, and uncontrolled development—everything that is now classified as ecology—had also begun to attract attention. There were plenty of reasons to call science into question!

Cartier described how Grothendieck, swept with these movements, even exasperated those who shared his political views and ideals by using excessive tactics. Grothendieck hit a couple of French policemen in Nice in September 1970 and was arrested, but quickly released when the authorities found out that he was a professor; they did not want to antagonize academia. Cartier writes:[13]

> His route is quite close to that of Simone Weil, and political anarchism took on more and more a religious tone in him. But, while Simone Weil's Catholicism was violently anti-Semitic (in 1942!), Grothendieck's Buddhism bears a strong resemblance to the practices of his Hasidic ancestors. For a long time he was receptive to all sorts of marginal "hippies," which resulted in his indictment and an absurd trial in 1977 due to a 1945 regulation that made it a misdemeanor to meet with a foreigner. He enjoyed playing the role of a modern

> Socrates, and was given a suspended sentence of six months in prison and a fine of 20,000 francs. It seems to me that his definitive break with science dates from this incident. He withdrew more and more into his own tent.

Grothendieck's leaving mathematics was related to his falling out with Bourbaki. Already in the 1957 Bourbaki congress, Grothendieck had begun to have doubts about continuing to work with Bourbaki, and even doubts about continuing to do mathematics.[14] Even at this early stage he felt strongly that Bourbaki should redo its work on the foundations of mathematics, changing these from set theory to category theory. In addition to its great generality and power, category theory does not suffer from the inherent limitations of set theory. As Pierre Cartier puts it: "Set theory is too restrictive: an element is either a member of a set or not, there is nothing in between."[15] And then, of course, there are the great paradoxes in set theory, which make the discipline full of theoretical holes.

Some members of Bourbaki supported Grothendieck's push toward category theory. But others—and these were the influential ones—felt that the foundation had already been laid and there was no going back. Grothendieck tried hard to convince them of the importance of this new proposed direction for the group. But he failed miserably. Grothendieck then got up, left in anger, and never came back to Bourbaki again. His own writings at the IHES had already begun to supplant the entrenched and inflexible work of Bourbaki.

In his memoirs, Grothendieck described his disagreement with Bourbaki, as well as his views about the group, its mission and how it operated, and the vision of structure in mathematics. He wrote as follows:[16]

> I am struck by the fact that I haven't here thought of the vast synthesis of contemporary mathematics that attempted to present the (collective) treatise of N. Bourbaki. This happened, it seems to me, for two reasons. First, this synthesis dealt with "putting in order" a vast collection of ideas and results that had already been known, without bringing into it new thought. If there is in all this a new idea, it is that of a precise mathematical definition of "structure," which revealed itself as the precious thread that runs through the entire treatise. But this idea seems to me similar to that of an intelligent and imaginative lexicography, rather than a new element of language, providing a new understanding of reality (here, mathematical reality). Additionally, since the 1950s, the idea of structure has become passé, superseded by the influx of new "categorical" methods in certain of the most dynamical areas of mathematics, such as topology or algebraic geometry. (Thus, the notion of "topos" refuses to enter into the "Bourbaki sack" of structures, decidedly already too full!) In making this decision, in full cognizance, not to engage in this revision, Bourbaki has itself renounced its initial ambition, which had been to furnish both the foundations and the basic language for all of modern mathematics.

Of course, Grothendieck is the inventor of topos as well as other mathematical ideas that were not incorporated by Bourbaki. It is easy to understand his feelings about Bourbaki's actions, and to understand why he left the group. History would prove him right: Bourbaki would decline following its refusal to accept new methods.

IN 1970, GROTHENDIECK discovered that government money was being used to support the IHES, and that some of this money was coming from military sources. This caused a great crisis in Grothendieck's life, and as a result he summarily quit his professorship in the institute. With the help of Jean-Pierre Serre, who held a professorship at the Collège de France, Grothendieck obtained a two-year temporary professorship at the Collège de France in Paris. But by now his interests had completely changed, from mathematics to ecology. He had founded the environmental group Survivre et Vivre, and was very active in political circles of the environment and antiwar action.

In order to obtain the professorship at the Collège de France, Grothendieck needed to give the administration of the Collège a syllabus for each of the courses he would teach. But he didn't want to lecture on mathematics—he only wanted to talk about the environment and to speak against nuclear weapons and the war in Vietnam. This was not acceptable to the administration, and his position was not renewed beyond 1973.

Grothendieck then accepted a professorship at the University of Montpellier, and taught there until 1984. His memoirs make it clear that he was quite depressed to teach there, in the backwater of French academic life, but he had no other possibility. His infatuation with political causes had made him a great liability to his friends and colleagues. Nobody could help him find a better position. Discouraged by teaching, Grothendieck applied for a research position with the French national center for scientific research (CNRS), the employer of most academic research scientists in France. But this research post did not work our for him, either.

From 1983 to 1985, he wrote his memoirs, *Recoltes et Semailles*, a work that remains unpublished. It includes personal recollections as well as mathematical discussions.

Grothendieck lived in the countryside near Montpellier, and spent most of his time writing thousands of pages of meditations on various topics. Many of these writings he would later burn.

In 1988, he was awarded the Crawford Prize in mathematics jointly with his former student, Pierre Deligne. He refused the prize. On January 10, 1988, Grothendieck retired officially from all his positions.[17] In August 1991 he suddenly left his home and disappeared into the Pyrenees, where he is believed to be hiding from the world. Several people have attempted to find him over the years, but he remains elusive. People who managed to see him have reported that he is obsessed with the devil. Grothendieck believes that the devil is constantly working to destroy the harmony of the world. Among the many bad things the devil has done are the creation of pollution, the destruction of the environment, and the promotion of war and destruction.

When he was last seen ten years ago, Grothendieck was obsessed with the meter. Throughout his life, he shunned physics and related sciences, seeing in them instruments of evil: physics brought us the nuclear bomb and other bad things. In his isolation, Grothendieck apparently began to contemplate the meter, and he came to the conclusion that here, too, the devil is at play. The devil, he believes, has maliciously replaced the nice whole number 300,000 km/second by the ugly number 299,887 km/second as the speed of light.[18]

What are the reasons for Grothendieck's decline and separation from society? Pictures from his younger days show him hugging his children lovingly—and yet these same children have not heard a word from him in years. Why would a man escape the world so completely that he does not interact with anyone anymore—even the people he once loved?

Those who have visited him caution against labeling him as mentally disturbed or even crazy. "Grothendieck is obsessed with evil," they say. "You must understand that about him—he fears the devil's work on Earth."[19]

Perhaps the answer lies in Grothendieck's childhood. He was deprived of every human decency growing up in the camps of wartime France. Here was a boy with incredible mathematical ability, and yet the teachers in these camps—who gave him the very little education he did receive—could not provide him with answers, challenges, or even encouragement. The boy grew up seeing the evil in the world from a very young age. He witnessed human brutality as few have (other than children who grew up in the Nazi death camps).

On his own, and with his incredible ambition, he taught himself mathematics and became one of the greatest mathematicians of all time—a man who could create mathematics rather than just contribute a few results. But he wanted more. He wanted to use his brain to improve the world, to rid it of the work of the devil, as it were. And here, predictably, he failed. For while the structural approaches in mathematics and science can lead to great results, society itself is not easily cured of its great ills.

Grothendieck could not succeed with Bourbaki; he was simply way above the group's modest abilities to innovate in mathematics. And once he left mathematics, he failed as a leader of environmental and antiwar causes. The brain that could guide him so well in mathematics could not supply him with the ability to effect political change. For, realistically, what can one man—working alone or with a group of followers—do to stop the cold war? Stopping the cold war was outside the realm of the possible not only for Grothendieck but for virtually everyone else on earth.

It was, perhaps, this foray into the area of political activity that doomed Grothendieck more than anything else. Virtually all the mathematicians of Bourbaki were politically active. All of them supported environmental causes and were strongly against all wars. In France—more than in America or other countries—academics tend to be overwhelmingly politically active, and almost always are on the side of social causes and are antigovernment and antiwar.

But these mathematicians did their "duty." Some of them even went to Vietnam to protest the war and others took different political actions—then they returned to mathematics, or to other activities. For some reason—probably having to do with the unfortunate circumstances in which he grew up—Grothendieck simply could not do that. Once he left mathematics, it was forever. So once he failed in his political activities, and by then had burned all his academic bridges, Grothendieck had nothing left to do. He was disappointed with Bourbaki, and he was disappointed with his own failure as a political activist. He blamed the devil for all the wrongs in the world, and had to withdraw from this imperfect world. He found the only place he could where pollution is minimal, where people—the few he meets—are peaceful, and where he does not hear what governments are doing. Grothendieck withdrew into the Pyrenees.

fifteen

THE DEATH OF BOURBAKI AND HIS LEGACY

THE OULIPO WRITER and mathematician Raymond Queneau had this to say about Bourbaki in 1962[1]

> *Il a nécessairement vieilli, votre fictif*
> *mathematician, il doit avoir pris du retard.*
> *Eh bien! Non, Bourbaki n'a pas vieilli*
> *parce qu'il ne* peut pas *vieiilir.*
>
> Translation: "He must have gotten old, your fictitious mathematician, he had to have surrendered to time. Well, no. Bourbaki has not gotten old because he *cannot* get old."

Bourbaki can't get old because he never existed. But many mathematicians believe that Bourbaki is not old—he is dead. The reasons for this belief are that Bourbaki has not published in many years; he exerted his influence on mathematics in the past, but does not count as a player in the field today; and finally, of the official members in Bourbaki today, none is among the highest ranked French mathematicians. Bourbaki, for all intents and purposes, is dead.

Christian Houzel, who was president of the French Mathematical Society, wrote:

> The age of Bourbaki and fundamental structures is over. I cannot say to what extent this is conditioned by the internal dynamics of the development of mathematics, or by ideological currents like the degradation of science's superior image in public opinion and scientists' questioning of the social status of their practice.[2]

WHY DID BOURBAKI and its ideas decline after the 1970s? What are some of the new approaches to mathematics that are now sweeping the academic world, and why do these ideas generally fare better than Bourbaki's? Which of Bourbaki's ideas are still dominant today? And why do these ideas survive?

The decade of the 1970s witnessed the decline of both Bourbaki on the one hand and French structuralism on the other. At the same time, Bourbaki also gave up its role as the cultural connector among the various fields of human study in which structuralism was important.[3]

Bourbaki achieved its goals of axiomatizing mathematics, stressing structure, and promoting rigor in a discipline that had fallen into laxity in the decades before the emergence of the group. Bourbaki began its decline after the end of the 1960s precisely because it had achieved its goals so marvelously. Henceforth, mathematics was carried out in a much more rigorous and precise way than had been done prior to the group's great contributions. There was, therefore, no more need for the group—there was nothing more to innovate.

Another factor that contributed to the decline of Bourbaki was that its members became well-known mathematicians under their own names. They were being awarded prizes and medals, the same medals they themselves had fought so strongly against in the early days of the group. The Bourbaki organization had become too powerful; and yet it now lacked its main purpose since all its goals had by now been achieved.

Bourbaki lost an incredibly important opportunity to remake its oeuvre in the new form of the theory of categories, something that would have better suited the study of structures than did the old theory of sets with its myriad problems and inadequacies. In addition, new theories in mathematics, such as chaos and fractals, as well as René Thom's catastrophe theory, emerged and demonstrated that structuralism is not absolutely necessary for doing good mathematics.[4]

Bourbaki had a chance, through the work of Grothendieck and his students, to refound modern mathematics on the theory of categories, but Bourbaki missed that chance. In part, this missed opportunity led to the demise of Bourbaki. For mathematics remained based on a flawed system, set theory, rather than something that would have been much more appropriate. Bourbaki had a possibility in the late 1960s to redirect itself toward a more ambitious goal. Mathematics had evolved further and reached a place in which new foundations could be laid for the discipline. This direction would have been possible because of the work of one man: Alexandre Grothendieck.

But a rift materialized in the midst of the group, and the members of Bourbaki could not agree to follow in the new direction. Grothendieck left the group in anger. Eventually, as we have seen, he would completely withdraw from society.

There also followed a serious battle for copyright with the group's publisher, a fight that sapped all of Bourbaki's energies over several years and left it ineffective and ineffectual. The legal battle began in 1975, when Bourbaki sued its publisher, Hermann, for the copyright of its books and the right to translate and publish them abroad. The group hired an excellent attorney, Blausteil, who had been the attorney for the heirs of Picasso. The matter was legally settled in 1980, and Bourbaki won. The group was now allowed to sign a publication deal with a new publisher. The members of Bourbaki could now republish all their older texts, and much work went into revising these texts, which left little time and energy for writing new books. Bourbaki had thus won the battle, but lost the war. This legal battle left it weary, exhausted, low on funds, and internally divided. It was left with no aim and no steam.[5]

Thus Bourbaki's pace was slowed down considerably, precisely because the group did win the legal battle for the older books. There followed several failed projects, such as a treatise on functions of several complex variables. Having lost its momentum, the group began its slide toward oblivion.

MOST MATHEMATICIANS AGREE that Bourbaki is dead. None of its present members are among the top forty or so mathematicians in France; the group does not publish; and mathematicians in general do not consider the group "alive." The questions that remain are when Bourbaki died, and why.

One theory is that the student revolt that swept France in 1968, affecting all of society, politicizing people whose political agendas had until then remained latent, and bringing the entire economy to a standstill, spelled the end of Bourbaki. According to this theory, Grothendieck and many other mathematicians in France took to politics as a more important occupation than doing mathematics and thereafter, mathematics was not the same.

Pierre Cartier, on the other hand, even though he took part in the political upheavals of the 1960s and 1970s and even went to Vietnam, as did Grothendieck, to protest against the war, believes that Bourbaki lived throughout most of the twentieth century. And, as he puts it, the twentieth century can be viewed as starting and ending with Sarajevo; so Bourbaki died with the war in Yugoslavia. Certainly Bourbaki was dead when the Berlin Wall came tumbling down toward the end of the century.[6]

These views tie together Bourbaki's work in mathematics and the group's political activity and agendas. This, perhaps, is a good description of Bourbaki's life and death, since the group has always been politically active, as individuals as well as collectively. But what else contributed to its demise? Personalities and their interactions played a major role in the dissolving of the Bourbaki group. André Weil was the main founder of Bourbaki, and its most active member. But Weil left France permanently during the war, and his non-presence in France made things difficult for the group. Following his departure, Bourbaki lacked a leader and a key figure.

It is true that Weil returned to France sporadically, and attended most of the congresses after the war, but he was not in Europe day-to-day, hence he was unable to participate in many activities that took place outside of the formal meetings. In those days, the 1960s and 1970s, communication across the ocean was not as easy and free-flowing as it is today, with fax machines and the Internet connecting people in real time. Long-distance phone calls, the only real-time connection in those days, were very expensive. In addition, there was the psychological factor of Weil's separation from the group.

Bourbaki tried to bridge this gap by listing Nicolas Bourbaki's official affiliation as the University of Nancago. Nancago was the fusion of Nancy, where some of the members were on the faculty, and Chicago, where André Weil was at that time. But this didn't help very much.

Additionally, as is the case with every organization, there was a danger that the members would pursue personal goals rather than the common goals of the group. Or rather, in the case of Bourbaki, the pertinent question is: How much effort does a member devote to the common goals, as compared with the efforts he devotes to personal goals? This question is especially important when applied to the leader of the group. In the case of André Weil, it is very likely that personal goals dictated the agenda. Weil remained in the United States for personal reasons even when the threat of war was over and while clearly Bourbaki needed him in France.

This was in stark contrast with the actions of another member, Claude Chevalley, who indeed returned to France and resumed his activities within the group. From his own memoirs, one gets the impression that Weil cared about his personal life more than he did the common good.

The conflict between personal goals and common goals clearly had its effect and swept with it the entire group of Bourbaki. We know, for example, that Jean Dieudonné did most of the writing of the final drafts of the books. Such an effort required great discipline and putting aside personal goals. For, however we may view the situation, a group that takes only collective credit for its works cannot last very long.

Everyone, at some point, wants to go it alone and get all the glory that's there to be had. Perhaps a comparison with the breakup of the Beatles is not out of place here. And the conflict between personal goals and common ones is most evident with people who are able to do great things. Mathematicians who can prove great theorems and win Fields Medals may at some point feel they are wasting their time doing thankless work for the common good. And perhaps this is exactly the reason why today the "members of Bourbaki," if they can still be defined that way, are not numbered among the better mathematicians in France.

It seems that Weil himself never considered his founding of Bourbaki the greatest achievement in his life. In the last page of his memoirs, Weil summarizes his important accomplishments. Here, he writes about travel and what travel has meant to him. He repeats the places to which he had traveled, including India, China, and Italy, and reminisces about how wonderful it had been to travel to these places, mostly in the company of his wife, Eveline. Nowhere in this summary of his life does he mention his cofounding of the Bourbaki group.[7]

From other information in his autobiography, one gets the distinct impression that Weil was infatuated with the childish pranks of "inventing" a person who never existed, creating for him false papers and a false identity, complete with a daughter, Betti, who even gets married, parents and relatives, and membership in a nonexistent Academy of Sciences of the nonexistent nation of Polvedia.

Weil was so taken with these activities that he even listed, as his only honor by the time of his death "Member, Poldevian Academy of Sciences." It seems that Weil could simply not go beyond these games: he could not grasp the deep significance and power of the organization he helped found. He was too close, and thus unable to see the great achievements Bourbaki was producing and to acknowledge and promote these achievements. Bourbaki changed the way we do mathematics, but Weil really saw only the pranks and the creation of a nonexistent person.

Another factor that has led to the decline of Bourbaki has been the great opposition that the group has encountered in its work. Bourbaki has brought rigor, abstraction, generality, preciseness, and structure into modern mathematics. Most mathematicians have welcomed these developments. However, there came a point at which many mathematicians felt that Bourbaki had gone too far.

In reviewing the works of the group, mathematicians outside it have noted an overreliance on generality and abstraction. Bourbaki seems to many to have reached the point that the generality is more important than specific cases. Besides the fact that this trend makes Bourbaki's writings very hard to follow—and the group itself acknowledges that its books could not and should not be used as textbooks for teaching mathematics—this trend is contrary to intuition and to the way mathematics is actually done. In general, the human mind does not work in great generalities.

Rather, the progression is from the specific to the general. A mathematician would usually start working on a specific problem or theorem, and once it had been solved, move on to see if it can be generalized. Generalities mean more power, greater meaning, and far enhanced significance. But few start with a general statement.

Another problem is the abstraction and rigor. Abstraction and rigor in mathematics are necessary to ensure that results are precise and correct and that there are no missing or false steps in proofs. This is, of course, a very important aspect of mathematics, and its promotion is certainly something that Bourbaki should be commended for. However, the abstraction and rigor should be the tool, not the purpose. In Bourbaki's works, it often seems that the writers have turned abstraction into a goal, and rigor takes over and leaves absolutely no place for intuition or even general understanding. Bourbaki does not, in general, use pictures or other visual aids: thus it completely discourages any understanding of the material that is "human." For how many people can see what is going on in a proof simply by following very technical details? Most mathematicians rely, at some point, on some kind of a mental picture of what is going on in a theorem or a proof.

It is this blind reliance on technical details, strict adherence to rigorous procedures, and an over appreciation of generalities at the expense of the specific case, the picture, the intuition, the human idea of a mathematical problem which have made Bourbaki disliked by some mathematicians. Having brought us their ideas of abstraction and generality and structure, Bourbaki lost its lead as the world of mathematics moved forward.

THE IDEAS OF French philosopher Michel Serres demonstrate the problems with Bourbaki's structures. In 1961, Serres realized, as did Piaget, that the idea of structure emanated directly from the mathematics of Bourbaki.[8] He defined structure mathematically as follows: "A structure is an operational set with an undefined meaning . . . grouping any number of elements, whose content is not specified, and a finite number of relations whose nature is not specified."[9] Serres noted that in algebra this definition of structure needs no explanation, and that in mathematics, therefore, the definition of structure is the truest.

But soon Serres diverged from the point of view that structures are important. Philosophy required methods and concepts that were far suppler than the rigid structures. He came to the belief that reality was chaotic rather than perfectly structured, and that in the sea of chaos that is reality there may be islands of rationality and structure. Knowledge did not need to be restricted, Serres claimed, and therefore Bourbaki's structuralism should give way to more general methods.

But the main reason for the decline of Bourbaki was that Bourbaki had achieved its goals and there was no longer a need for the group. Mathematics had entered the realm of axiomatic and structural precision. Proofs of theorems were now correct, and rigor became the name of the game. Mathematics could now proceed on its own route; it did not need to be guided by Bourbaki.

And finally, another reason for the decline of Bourbaki is the present decline of nationalism and chauvinism. Bourbaki was established as a *French* group—even though it always had foreign members, such as Armand Borel. Bourbaki did French mathematics, and its contributions to structuralism were also in a French context—through the works of Claude Lévi-Strauss and other French intellectuals. Today, with the general decline in nationalism and nationalistic feelings, there really is no place for a nationalistic mathematical group. Mathematics today is an international endeavor. Mathematicians travel the world, meeting in various countries, where cooperation across national boundaries is the norm. With the formation of a unified Europe with no national boundaries and a common currency as well as a parliament, there is no meaning anymore to a "French" mathematical group, or a "French" way of doing mathematics.

Mathematics—like science, philosophy, or the humanities—is now pursued by large international groups of participants.

A mathematical project may entail work on certain aspects of a theory by a French mathematician, a Japanese one, an Englishwoman, a Dutchman, and so on. Who, in fact, even looks at the national origin of a mathematician working in a group on some topic? So the age of a French—and completely male—group of mathematicians is simply over. Who wants to belong to a national or gender-specific group doing anything nowadays?

IF HE IS alive, Grothendieck is still hiding in the Pyrenees. He is hiding very well now, since attempts to find him have failed. Apparently, this is what he wants: to be alone, to write and destroy his own writing, and not to have any connection with people other than grocery store clerks or laborers who might do occasional work for him.

His disappearance and his anger with the world symbolize the demise of Bourbaki. For Grothendieck alone held the great hope for the future of Bourbaki. Grothendieck and his work were the next stage in the program of abstraction and generalization in mathematics that Bourbaki had embarked upon. Alexandre Grothendieck was the human incarnation of the essence of Bourbaki, of the ideals that Bourbaki strove for in mathematics, for here was a man who actually *thought* in great generalities, and for whom axiomatic thinking was natural. Grothendieck's oeuvre was, in fact, all about *structure*, so that the structuralist idealism of Bourbaki found in Grothendieck's work its finest manifestation.

But the man who left mathematics for political causes came up completely empty-handed. Fellow mathematicians were disappointed with his loss of interest in the discipline, and they saw that his political actions were ineffectual and completely useless. Realizing his own failure, Grothendieck drew further apart from the world around him. Perhaps reflecting his own parents' disillusionment with revolution following the defeat of the Republicans in the Spanish Civil War, Grothendieck realized that he could not change society with his political activity. He therefore chose to withdraw from society.

The man who could bring the most beautiful structures into mathematics could not bring structure to political reality. And perhaps politics, unlike other disciplines, is not amenable to the methods of structuralism. At any rate, political change was not to be brought on through the work of Grothendieck, and as the need for political change diminished, the man withdrew from society altogether. It is hoped that he enjoys the peaceful and pristine environment in which he now lives.

SELECTED BIBLIOGRAPHY

Aczel, Amir D. *Entanglement: The Greatest Mystery in Physics*. New York: Plume, 2003.

———. *Fermat's Last Theorem*. New York: Dell, 1997.

———. *God's Equation*. New York: Dell, 2000.

———. *The Mystery of the Aleph: Mathematics, Kabbalah, and the Search for Infinity*. New York: Pocket Books, 2001.

Aubin, D. "The Withering Immortality of Nicolas Bourbaki: A Cultural Connector at the Confluence of Mathematics, Structuralism, and the Oulipo in France." *Science in Context* 10, no. 2 (1997).

Badiou, Alain. Interview with Catherine Grenier. In *Big Bang: Destruction and Creation in Twentieth Century Art*. Paris: Centre Pompidou (2005).

Beaulieu, L. "A Parisian Café and Ten Proto-Bourbaki Meetings (1934–1935)." *The Mathematical Intelligencer* 15, no. 1 (1993).

———. "Proofs in Expository Writing—Some Examples from Bourbaki's Early Drafts." *Interchange* 21, no. 2 (1990).

Benet's Reader's Encyclopedia. Edited by Bruce Murphy. 4th ed. New York: HarperCollins, 1996.

Boas, R. P. "Bourbaki and Me." *The Mathematical Intelligencer* 8, no. 4 (1986).

Borel, Armand. "Twenty-Five Years with Nicolas Bourbaki, (1949–1973)." *Notices of the American Mathematical Society*, March 1989.

Bourbaki, Nicolas. *Théorie des ensembles*, Paris: Hermann, 1938.

Cartan, Henri. "Sur Jean Delsarte," in "Hommage à Jean Delsarte." *Nichifutsu Bunka* 25 (1970).

Cartier, Pierre. "A Mad Day's Work: From Grothendieck to Connes and Kontsevich, the Evolution of Concepts of Space and Symmetry." *Bulletin of the American Mathematical Society* 38, no. 4 (October 2001).

Chouchan, Michèle. *Nicolas Bourbaki: Faits et legends*. Paris: Editions du Choix, 1995.

Cox, Neil. *Cubism*. London: Phaidon, 2000.

Dolgachev, Igor. "Review of *The Geometry of Schemes* by David Eisenbud and Joe Harris." *Bulletin of the American Mathematical Society* 38, no. 4 (2001).

Dosse, François. *Histoire du structuralisme*. Paris: Editions La Découverte, 1992.

Gaultier, Alyse. *L'abécédaire du cubisme*. Paris: Flammarion, 2002.

Grenier, Catherine. "Destruction." In *Big Bang: Destruction and Creation in Twentieth Century Art*. Paris: Centre Pompidou (2006).

Grothendieck, Alexandre. "Promenade à travers une oeuvre ou l'enfant et la mère." Manuscript, 1986.

——— and J.-P. Serre. *Correspondance Grothendieck-Serre*, Providence, RI: American Mathematical Society, 2003.

Guedj, Denis. "Nicolas Bourbaki, Collective Mathematician:

An Interview with Claude Chevalley." *The Mathematical Intelligencer* 7, no. 2 (1985).

Houzel, Christian. "Les mathématiciens retournent au concret." *La Recherche* 10 (1979).

Jackson, Allyn. "Comme appelé du Néant—As if Summoned from the Void: The Life of Alexandre Grothendieck." *Notices of the American Mathematical Society* 51, no. 9 (October 2003).

Jakobson, Roman. *Essays on General Linguistics*. Bloomington: University of Indiana Press, 1952.

Le Lionnais, François. "La Lipo." In *Oulipo: La littérature potentielle*. Paris: Gallimard, 1973.

Lescure, Jean. "Petite histoire de l'Oulipo." In *Oulipo: La littérature potentielle*. Paris: Gallimard, 1973.

Lévi-Strauss, Claude. *Anthropologie structurale*. Paris: Plon, 1958.

———. *Les structures élémentaires de la parenté*. Berlin: Mouton de Gruyter, 2002.

Lévi-Strauss, Claude, and Didier Eribon. *De près et de loin*. Paris: Odile Jacob, 1988.

Loschack, Pierre, and Leila Schneps. "A Brief Timeline for the Life of Alexandre Grothendieck (Which Has the Advantage of Accuracy)." Manuscript. Paris, 2006.

Mashaal, Pierre. "La sage d'un nom." *Pour La Science* 2 (February-May 2000): 18.

———. "Le vrai general Bourbaki (1816–1897)." *Pour La Science* 2 (February-May 2000): 17.

Piaget, Jean. *Structuralism*. New York: Basic Books, 1970. Quoted in Aubin, p. 317.

Prague Circle, *The Theses of 1929* (republished as *Change*, Paris: Seuil, 1969)

Scharlau, Winfried. "Materialien zu einer Biographie von Alexander Grothendieck." Manuscript, 2004.

Schwartz, Laurent. *Un mathématicien aux prises avec le siècle.* Paris: Odile Jacob, 1997.

Senechal, Marjorie. "The Continuing Silence of Bourbaki—An Interview with Pierre Cartier, June 18, 1997." *The Mathematical Intelligencer* 1 (1998).

Serres, Michel. *Hermès I: La communication.* Paris: Minuit, 1968.

Treves, François. "Biographical Sketch of Laurent Schwartz (1915–2002)." *Notices of the American Mathematical Society* 50, no. 9 (October 2003).

Weil, André. *The Appenticeship of a Mathematician* Translated by Jennifer Gage. Boston: Birkhäuser, 2002.

NOTES

Chapter 1

1. Winfried Scharlau, "Materialien zu einer Biographie von Alexander Grothendieck" (manuscript, 2004), 21–2. According to Scharlau, who has studied Grothendieck's parents' history, there are two other possible dates for Sacha Shapiro's birth, depending on the source one consults. One date is August 8, 1890, and the other is November 10, 1889.
2. Ibid., 35.
3. Ibid., 12.
4. Allyn Jackson, "Comme appelé du Néant—As if Summoned from the Void: The Life of Alexandre Grothendieck," *Notices of the American Mathematical Society* 51, no. 9 (October 2003): 1040.
5. Scharlau, "Materialien zu einer Biographie," 19.
6. Ibid., 53.
7. Jackson, "Comme appelé du Néant," 1040.
8. Scharlau, 76 (translated by the author).
9. Alexandre Grothendieck, "Promenade à travers une oeuvre ou l'enfant et la mère" (manuscript, 1986): 1.
10. Ibid.
11. Pierre Cartier, "A Mad Day's Work: From Grothendieck to Connes and Kontsevich, the Evolution of Concepts of Space

and Symmetry," *Bulletin of the American Mathematical Society* 38, no. 4 (October 2001): 391.

12. Grothendieck, "Promenade," 1.
13. Ibid., 2.
14. Scharlau, "Materialien zu einer Biographie," 81.
15. I am indebted to the French mathematicians Pierre Loschak and Leila Schneps for this information.
16. Cartier, "A Mad Day's Work," 391.
17. Grothendieck, "Promenade," 2.
18. Ibid., 3.
19. Pierre Cartier, personal communication.

Chapter 2

1. André Weil, *Apprenticeship,* 129.
2. Ibid., 130.
3. Ibid.
4. Ibid., 131.
5. Ibid., 133.
6. Ibid., 134.
7. Ibid., 15.
8. Ibid., 18.
9. Ibid., 27.
10. For a description of this event see Amir D. Aczel, *God's Equation* (New York: Dell, 2000).
11. Pierre Cartier, personal communication.
12. Weil, *Apprenticeship,* 40.
13. Ibid.
14. Ibid.
15. Ibid., 41.
16. Ibid., 42.
17. Ibid., 54.
18. Ibid., 55.
19. Ibid., 95.
20. Ibid., 53.
21. Ibid., 56.
22. Ibid., 66.

23. Ibid., 82.
24. Ibid., 134.
25. Ibid., 152.

Chapter 3

1. François Treves, "Biographical Sketch of Laurent Schwartz (1915–2002)," *Notices of the American Mathematical Society* 50, no. 9 (October 2003): 1072–8.
2. I am indebted to the mathematician Leila Schneps in Paris for this story.
3. Laurent Schwartz, *Un mathématicien aux prises avec le siècle* (Paris: Odile Jacob, 1997): 207.
4. Weil, *Apprenticeship,* 169–170.
5. Ibid. 170.

Chapter 4

1. Grothendieck, "Promenade," 4.
2. Ibid., 2.
3. Jackson, "Comme appelé du Néant," 1042.
4. Grothendieck, "Promenade," 5 (translated by the author).
5. Ibid., 6 (translated by the author).
6. Ibid.
7. Ibid., 6–7 (translated by the author).
8. Grothendieck, "Promenade," 15 (n. 17).
9. Jackson, "Comme appelé du Néant," 1039.
10. Ibid., 1043.

Chapter 5

1. Pierre Mashaal, "Le vrai general Bourbaki (1816–1897)," *Pour La Science* 2 (February-May 2000): 17.
2. Pierre Mashaal, "La sage d'un nom," *Pour La Science* 2 (February-May 2000): 18.

3. For the story of the quantum revolution see Amir D. Aczel, *Entanglement: The Greatest Mystery in Physics*, New York: Plume, 2003.
4. For the complete story of Georg Cantor and his work on set theory, see Amir D. Aczel, *The Mystery of the Aleph: Mathematics, Kabbalah, and the Search for Infinity* (New York: Pocket Books, 2001).

Chapter 6

1. Neil Cox, *Cubism* (London: Phaidon, 2000): 31.
2. Ibid., 35.
3. Ibid., 36.
4. Catherine Grenier, "Destruction," in *Big Bang: Destruction and Creation in Twentieth Century Art* (Paris: Centre Pompidou, 2005): 41.
5. Alain Badiou, in an interview with Catherine Grenier, in *Big Bang: Destruction and Creation in Twentieth Century Art* (Paris: Centre Pompidou, 2005): 35.
6. Alyse Gaultier, *L'abécédaire du cubisme* (Paris: Flammarion, 2002): 97.
7. Ibid., 94.

Chapter 7

1. André Weil, *Apprenticeship* , 97.
2. Pierre Cartier, personal communication.
3. L. Beaulieu, "A Parisian Café and Ten Proto-Bourbaki Meetings (1934–1935)," *The Mathematical Intelligencer* 15, no. 1 (1993): 27–35.
4. Ibid.
5. Henri Cartan, "Sur Jean Delsarte," in "Hommage à Jean Delsarte," *Nichifutsu Bunka* 25 (1970): 27–30.
6. Beaulieu, "A Parisian Café," 29.
7. Ibid., 30.
8. Ibid.
9. Pierre Cartier, in discussion with the author, June 21, 2005.
10. Beaulieu, "A Parisian Café," 31.
11. Ibid.

12. Ibid., 32.
13. Ibid. According to Beaulieu, the average person in this group had by that time published twenty-three research papers and the numbers varied from four to seventy-five.
14. Denis Guedj, "Nicolas Bourbaki, Collective Mathematician: An Interview with Claude Chevalley," *The Mathematical Intelligencer* 7, no. 2 (1985): 19.
15. Reported in Beaulieu, "A Parisian Café," 33.
16. Weil, *Apprenticeship,* 101.
17. Ibid., 112.
18. Guedj, "Collective Mathematician," 20.
19. Ibid.
20. Weil, *Apprenticeship,* 113.
21. Pierre Cartier, personal communication.
22. Guedj, "Collective Mathematician," 20.
23. Ibid.
24. Ibid. 20.
25. Weil, *Apprenticeship,* 114.
26. Ibid., 121.
27. Ibid.,127.
28. Ibid., 124.
29. Ibid.

Chapter 8

1. According to the traditional history of Bourbaki, there were six original founders. Szolem Mandelbrojt joined the six later. Some members joined and others left, so reports vary as to the exact number of founders and of members at any given point in time.
2. Guedj, "Collective Mathematician," 18.
3. ibid.
4. ibid.
5. Ibid., 22.
6. Marjorie Senechal, "The Continuing Silence of Bourbaki—An Interview with Pierre Cartier, June 18, 1997," *The Mathematical Intelligencer* 1 (1998): 22–8.
7. Pierre Cartier, in discussion with the author, June 21, 2005.

8. Senechal, "Continuing Silence," 22.
9. Ibid.
10. Ibid., 23.
11. Ibid.
12. Ibid.
13. Pierre Cartier, personal communication. When I attended the Bourbaki Seminar in Paris in June 2005, the lecture hall was not full; a mathematician was writing continuously on the board, talking about a problem in algebraic topology; people were taking notes. It did not seem like an important mathematical event; it was more like a minor talk at a modern mathematical meeting.

Chapter 9

1. Senechal, "Continuing Silence," 23.
2. Ibid., 24.
3. Armand Borel, "Twenty-Five Years with Nicolas Bourbaki, (1949–1973)," *Notices of the American Mathematical Society* (March 1989): 374.
4. Ibid.
5. *Correspondance Grothendieck-Serre,* Providence, RI: American Mathematical Society, 2003.
6. Borel, "Twenty-Five Years," 374.
7. Ibid.
8. Ibid., 375.
9. Guedj, "Collective Mathematician," 20.
10. Cartier, personal communication.
11. Guedj, "Collective Mathematician," 21.
12. Borel, "Twenty-Five Years," 375.
13. Ibid.
14. Ibid., 376.
15. Senechal, "Continuing Silence," 24.
16. Ibid.
17. Borel, "Twenty-Five Years," 376.
18. Ibid.
19. Senechal, "Continuing Silence," 25.

20. L. Beaulieu, "Proofs in Expository Writing—Some Examples from Bourbaki's Early Drafts," *Interchange* 21, no. 2 (1990): 36.
21. R. P. Boas, "Bourbaki and Me," *The Mathematical Intelligencer* 8, no. 4 (1986): 84.
22. Ibid.
23. Ibid.
24. Ibid.
25. Ibid.
26. Pierre Cartier, in discussion with the author, June 21, 2005.
27. François Dosse, *Histoire du structuralisme* (Paris: Editions La Découverte, 1992): 260.
28. Ibid., 335–6.

Chapter 10

1. *Benet's Reader's Encyclopedia*. Edited by Bruce Murphy. 4th ed. (New York: HarperCollins): 990.
2. D. Aubin, "The Withering Immortality of Nicolas Bourbaki: A Cultural Connector at the Confluence of Mathematics, Structuralism, and the Oulipo in France," *Science in Context* 10, no. 2 (1997): 309.
3. Ibid.
4. Dosse, *Histoire du structuralisme*, 26.
5. Ibid., 27.
6. Claude Lévi-Strauss and Didier Eribon, *De près et de loin* (Paris: Odile Jacob, 1988): 47.
7. Claude Lévi-Strauss, *Anthropologie structurale* (Paris: Plon, 1958): 46.
8. Ibid., 48.
9. Lévi-Strauss and Eribon, *De près et de loin*, 79.
10. Claude Lévi-Strauss, *Les structures élémentaires de la parenté*, "Appendix to Part I, by André Weil" (Berlin: Mouton de Gruyter, 2002): 258.
11. Dosse, *Histoire du structuralisme*, 29–30.
12. Aubin, "Withering Immortality," 301.
13. Ibid.
14. Ibid., 311.

15. Ibid., 314.
16. Ibid.
17. Ibid., 316.
18. Dosse, *Histoire du structuralisme*, 29.

Chapter 11

1. Ibid., 75.
2. Prague Circle, *The Theses of 1929* (republished as *Change*, Paris: Seuil, 1969): 31.
3. Dosse, *Histoire du structuralisme*, 76–7.
4. Ibid., 46.
5. Roman Jakobson, *Essays on General Linguistics* (Bloomington: University of Indiana Press, 1952).
6. Lévi-Strauss, *Structural Anthropology*, 79.
7. Dosse, *Histoire du structuralisme*, 95.
8. Ibid., 97–8.
9. Ibid., 100–1.

Chapter 12

1. Jean Piaget, *Structuralism* (New York: Basic Books, 1970): 17. Quoted in Aubin, 317.
2. Aubin, "Withering Immortality," 318.
3. Ibid., 317.
4. Ibid., 319.
5. Ibid.
6. Ibid.
7. Ibid., 315.
8. Dosse, *Histoire du structuralisme*, 132.

Chapter 13

1. Aubin, "Withering Immortality," 320.
2. Ibid., 322.

3. Ibid., 323.
4. Ibid., 297–342.
5. François Le Lionnais, "La Lipo," in Oulipo, *La littérature potentielle* (Paris: Gallimard, 1973):17–18.
6. Jean Lescure, "Petite histoire de l'Oulipo," in Oulipo, *La littérature potentielle* (Paris: Gallimard, 1973): 25.
7. Oulipo, *La littérature potentielle* (Paris: Gallimard, 1973): 233.
8. Ibid., 245.
9. Ibid., 87.
10. Ibid., 161 (translated by the author).
11. Ibid., 217, (translated by the author).

Chapter 14

1. Jackson, "Comme appelé du Néant," 1047.
2. Ibid.
3. Igor Dolgachev, Review of *The Geometry of Schemes* by David Eisenbud and Joe Harris, *Bulletin of the American Mathematical Society* 38, no. 4 (2001): 469.
4. Ibid.
5. Cartier, "Mad Day's Work," 393.
6. Ibid.
7. Ibid., 395.
8. Michèle Chouchan, *Nicolas Bourbaki: Faits et legends* (Paris: Editions du Choix, 1995): 33.
9. Jackson, "Comme appelé du Néant,"1050.
10. Ibid., 1051.
11. Ibid., 1038.
12. Cartier, "Mad Day's Work," 393.
13. Ibid.
14. Ibid., 392.
15. Pierre Cartier, in discussion with the author, June 21, 2005.
16. Grothendieck, "Promenade," 63, n. 78 (translated by the author).
17. This information is based on "A Brief Timeline for the Life of Alexandre Grothendieck (Which Has the Advantage of Accuracy)," by Pierre Lochak and Leila Schneps.

18. Cartier, "Mad Day's Work," 393.
19. Pierre Lochak, in discussion with the author, June 24, 2005.

Chapter 15

1. Aubin, "Withering Immortality," 297.
2. Ibid., 299.
3. Ibid., 323.
4. Ibid., 324.
5. Pierre Cartier, in discussion with the author, June 21, 2005.
6. Pierre Cartier, personal communication.
7. Weil, *Apprenticeship*, 117.
8. Aubin, "Withering Immortality," 324.
9. Michel Serres, *Hermès I: La communication* (Paris: Minuit, 1968): 32.

INDEX